COLLINS GEM
CATS
a mine of information

D0978059

COLLINS GEM
HORSES
& PONIES
a mine of information

COLLINS GEM
INSECTS
a mine of information

COLLINS GEM
KINGS &
QUEENS
a mine of information

COLLINS GEM
MUSHROOMS
& TOADSTOOLS
a mine of information

COLLINS GEM
SNAKES
a mine of information

COLLINS GEM
SPIDERS
a mine of information

COLLINS GEM
STRESS
Survival Guide
a mine of information

COLLINS GEM
TAROT
a mine of information

COLLINS GEM
WINE
Guide
a mine of information

COLLINS GEM
WORLD
atlas
a mine of information

COLLINS GEM
YOGA
a mine of information

COLLINS GEM
ZODIAC
Types
a mine of information

COLLINS Jane's

CIVIL AIRCRAFT

Richard Aboulafia

HarperCollins*Publishers*

Dedication:
To my parents, who taught me to write in the first place, and to the memory of John Miles III, a fine man who loved aeroplanes.

Thanks to the late Mr William Crampton of the Flag Institute for the flags

HarperCollins Publishers
PO Box, Glasgow G4 0NB

First published 1996
This edition published 1999

Reprint 10 9 8 7 6 5 4 3 2 1 0

ISBN 0 00 472264-7

Printed in Italy by Amadeus S.p.A.

Contents

General Aviation/ Business Aircraft

Helicopters

Introduction: Strength Through Overcapacity

All sectors of the civil aircraft market - jetliners, turboprop transports, business aircraft, and helicopters - have one unfortunate thing in common. There is a terrific diversity of models. This is great for people who fly (and watch) aircraft but terrible for people who build them. Skewed by government subsidies and outright jobs programmes, most civil aircraft markets have too many players to allow more than a few of them to make a profit.

Aircraft builders in the USA and Europe have also been hit by the terrific downturn in defence spending. Even when they are not profitable, military aircraft programmes provide new technologies. These programmes also help firms achieve a volume of business necessary to offer products at competitive prices. Worst of all, the first half of the 1990s has seen the largest simultaneous downturn in civil and military aerospace markets since World War II.

At the Bottom of the Jetliner Trough

The civil jetliner market has been particularly hard hit. Jetliner deliveries fell from an all-time high of 844 in 1991 to about 500 in 1995. This 'deliveries trough' is taking place just as numerous new jetliner models are arriving on the market. Boeing, Airbus, and McDonnell Douglas are all trying to recoup their investments on these programmes, but falling delivery

The McDonnell Douglas MD11 faces tough competition

rates mean shrinking revenues. Narrowbody planes have been hit hardest. Traditionally, the narrowbody jetliner market has been dominated by Boeing and McDonnell Douglas, but Airbus changed this in the mid 1980s when it introduced its A320, followed by the A321 and A319. Douglas's DC-9/MD-80 series, now in production as the MD-90, has suffered considerably from this competition and from inadequate investment. Boeing's 737 will soon see its third incarnation, as the 737-600/-700/-800. With the 737 and 757, Boeing still dominates the narrowbody market. However, these three families are being nibbled at by the bottom of the narrowbody market. Fokker's F100, the British Aerospace Avro RJ series, and McDonnell Douglas MD-95 all compete in the desperate 100-seater market, fighting with the A319 and 737-500/-600 for small orders. As Henry Kissinger said about academia, the competition is especially fierce when the stakes are so small.

Small jetliners are also affected by aviation industry Maxim Number One: if things can get worse, they will. Asian and European manufacturers would like to develop new 100-seat aircraft. Japan has its YSX 100-seat programme. Indonesia's IPTN has its N-2130 proposal. Korea and China want to build jetliners too. There is no market; it is just that this is the easiest place to start a jetliner industry. European manufacturers hope to take advantage of Asian aspirations by proposing new programmes that Asian manufacturers can pay to be part of.

Now the good news. Many narrowbody programmes will benefit from noise regulations, which will force noisy, Stage 2 aircraft (basically any narrowbody jetliners built before the early 1980s) out of service by the end of the century. However, airlines also have the option of re-engining their Stage 2 aircraft or of installing hush-kits. Re-engined 727s and hush-kitted DC-9s are proving fairly popular, particularly in the US. Some discount carriers, such as

Valujet, have used these cheap, upgraded aircraft to launch low-cost services in the middle of major airlines' route networks.

Widebody jetliners are about the only civil aircraft market segment not doomed to suffer from rampant overcapacity. They are also the most profitable segment, probably because the barriers to entry are very high. Only two countries - the US and former Soviet Union - have been able to make the leap to widebody aircraft manufacture. Europe only did it by pooling national resources into the Airbus consortium. This should serve as a warning to manufacturers seeking to develop a jetliner industry starting with a 100-seat aircraft: but it probably will not.

Unfortunately, McDonnell Douglas has failed to invest in all-new widebody designs, such as its proposed MD-XX 220-seater. The company is still building the MD-11, a DC-10 derivative, but this looks set to end production in the next few years. There will be no follow-on model. This will leave the widebody market to two players: Airbus with its A300/310 and A330/340, and Boeing with its 747, 767, and 777.

The 300-seat widebody segment is a promising growth market. The A330/340 (essentially the same aircraft with different propulsion philosophies) and

777 will compete here, replacing hundreds of DC-10s and Lockheed L-1011s. The combined order book for the A330/340 and 777 is approaching 500 aircraft, which is not bad for planes arriving in the middle of a recession.

Airbus and Boeing will also stretch their designs for high density routes and shorten them for long routes. Airbus will probably use a shortened A330 to replace the A300/310, while Boeing may use a shortened 777 to take over much of the 767 market.

The Next Big Thing

Beyond the current jetliners, attention is focused on new aircraft in two different flavours: bigger and faster. For years, ever-optimistic forecasts of world airline travel growth have led many observers to anticipate a new generation of huge super jumbo transports. In the early 1990s, the possibility emerged of a global consortium to build an all-new 600/800-seat leviathan. Manufacturers have since concluded that the market is not yet ready for a super jumbo. But Boeing plans to develop its legendary 747 into new stretched and re-winged versions, seating up to 600 passengers. Lacking the cash to develop a competing product, Airbus can only counter with a stretched A340 with only 370 seats.

In addition to super jumbos, some observers believe

The pioneering Boeing 747 inspired a generation of similar wide-bodied airliners and larger versions are planned

the future could bring a new supersonic transport to replace the Concorde. NASA, Japan's MITI, and other government agencies are paying for technology development efforts which could result in a new High Speed Civil Transport (HSCT). Of course, the companies that build planes are not spending any of their own money on HSCT development. They know that cost will always be the paramount factor in air transport, and that any jetliner which appeals to a limited number of rich travellers will be a market failure, just like the Concorde.

What the Former Soviets are Doing

The USSR had a world class jetliner industry, at least in terms of quantity. It produced a supersonic transport, the Tu-144, now out of service. It produced a single widebody transport, the Il-86. It produced numerous copies of Western designs such as the Tu-154, a 727 copy.

Unfortunately, none of the former Soviet jetliners can compete with Western models. The Tu-154 costs more to operate than the 727 and a lot more than the latest Western equivalents, the 757 or A321. Russian and other ex-Soviet airlines now have to pay for fuel, and care about operating costs. The formerly captive Eastern European airlines can now order what they want and are replacing their Tupolevs and Ilyushins with Boeings and Airbuses as fast as they can. As a response, former Soviet design bureaux are offering jetliners with Western engines and avionics. The Tu-204, with Rolls-Royce engines, and the Il-96, with Pratt & Whitney engines, could compete with Boeing, Airbus, and Douglas planes. These Westernized former Soviet aircraft may offer the best hope for former Soviet airlines, which are starved for hard currency. Yet they will probably only garner a few export sales, probably in traditional Russian aerospace markets. This means, India, China, and countries that want to pay for their aircraft with palm oil.

The first four Dornier Do.328s seen in formation

Regional Aircraft: A Great Way to Lose Money

If the jetliner market economics are twisted, then turboprop market economics are completely sprained. There are basically no barriers to market entry. As a result, the 30-seat regional aircraft segment alone has five different models - Saab's 340, DHC's Dash 8-100, Dornier's 328, British Aerospace's J41, and Embraer's EMB-120 - competing for market share. Deliveries of all of these are averaging fewer than 100 per year. The situation in the 20/40/50/60/70-seat segments is about as bad. Of course, none of these programmes can be considered truly profitable.

Profit-minded US companies seldom bother with regional aircraft. Only Fairchild and Raytheon's Beech unit offer turboprop transports. The 19-seat Beech 1900 takes advantage of commonality with the company's profitable King Air product line. Fairchild's Metro takes advantage of the US military market, which buys the 19-seat plane as the C-26 for transport duties. Fortunately, the world regional aircraft industry is beginning to show signs of sanity and consolidation. Only one player - Short Brothers - has actually left the market, after the 330/360 series ended production in 1991. But in early 1995, British Aerospace merged its regional aircraft programmes with Aerospatiale/Alenia's ATR product line creating the Aero International Regional (AIR) alliance. BAe agreed to kill its disastrous ATP/J61 programme as the price for this merger.

More mergers and product cancellations will take place in the future. If Fokker and Daimler Benz Aerospace merge their regional aircraft with AIR's, Europe could be close to a sane, rational regional aircraft family, especially if Saab and CASA join too. This merger would certainly kill the Fokker 50, a direct competitor of the ATR42. Just as existing regional aircraft players are rationalizing, new regional aircraft market entrants are coming on line, most

A Raytheon Hawker 800 over London's Docklands

notably Indonesia's IPTN. The IPTN N-250, the first Asian transport since Japan's YS-11, revives the prospect of further Asian aerospace ventures. Russia's Ilyushin wants to compete on the world market with a Westernized Il-114, and Antonov has a Westernized An-38. In short, there are more new regional aircraft programmes than expiring ones.

Meanwhile, the current generation of turboprop transports are menaced by the spectre of the turbofan. Airline passengers often gnash their teeth at the sight of a propeller, and, inevitably, every year or two a turboprop transport accident produces a safety scare.

Canadair's RJ is now being ordered and delivered in large numbers, and Embraer's EMB-145 will soon arrive. These planes are small, but they look like mini-jetliners, and passengers may find them more acceptable.

These 50-seat jets could displace the similarly sized ATR42, DHC Dash 8-300, Saab 2000, and other turboprop transports as commuter and feeder aircraft of choice. Canadair will also stretch its RJ into the 70-seat CRJ-X, which will threaten the ATR72 and other large turboprop designs.

This Year's Model

New models have been called the life-blood of the business aircraft market. Companies and other private aircraft users want the latest thing, not a product near the end of its (perceived) life cycle. As a result, over a dozen new business jets and countless variants of existing aircraft are entering the market this decade. The market is not actually growing very much, but at least it is an exciting place to do business.

New technologies, especially engines, are changing the high and low ends of the business aircraft market. At the high end, the Rolls-Royce/BMW BR700 turbofan is powering both Gulfstream's GV and Bombardier/Canadair's Global Express. These large aircraft carry up to 19 passengers, cost more than some jetliners, and can cross the Pacific Ocean non-stop. At the low end of the market, the Rolls-Royce/Williams International FJ44 turbofan is allowing small, cheap business jets to challenge traditional turboprop aircraft such as the Beech King Air. Cessna's CitationJet and Swearingen's SJ30 are the first FJ44-powered business jets, but Raytheon is planning its own FJ44-powered design, the Premier One. Others may follow.

Despite these technological trends which promise to transform the business aircraft market, observers should note that the last big trend in this market was a complete flop. In the late 1980s, designers created numerous high-tech turboprop designs using foreplanes, pusher engines, and composite materials. Only two planes actually entered production - Beech's Starship and Piaggio's Avanti. Both have failed. The Starship ended production after 53 were built, and Piaggio is treading water waiting for brighter days. Hopefully, the new high- and low-end business jets will get a warmer reception. Meanwhile, the middle of the business market is trying to move to higher ground. Learjet's new Model 60 will be the heaviest and most capable Learjet yet. Cessna's new Citation X will be the largest Citation and, at Mach 0.9, the fastest civil aircraft other than the Concorde. Israel Aircraft Industries' new Galaxy will use the Astra's wing with a larger fuselage and more powerful engines.

While the number of business aircraft models is growing, the number of companies in the market is contracting. Textron's Cessna unit has always been big, with its Citation and Caravan families. In the past, Cessna was only rivalled by Learjet, which is now a shadow of its former self. But recently, two other conglomerates have emerged to challenge Cessna for market dominance. Bombardier, with Learjet and Canadair, and Raytheon, with Beechjet and Hawker (British Aerospace's old 125 line) are also striving for critical mass. Of the remaining business aircraft makers, Gulfstream is essentially for sale and could be purchased by Cessna or Raytheon. Dassault, Pilatus, Socata, and Israel Aircraft Industries build business aircraft alongside other aircraft and aerospace products. Of the many prospective new business aircraft entrants which have emerged over the last decade, only Swearingen looks set to deliver any planes.

Why the Civil Helicopter Market Needs Another Cold War

The downturn in defence spending affects helicopters and their makers more than any other civil aircraft segment. Civil helicopters are often variants of

The maiden flight of the Learjet 60, 13 June 1991

military designs. The Bell 206/407 series is based on the US Army OH-58, and Eurocopter's BO 105 was built initially as an anti-tank model for the German army. Even purpose-built civil machines use technology developed for military programmes. McDonnell Douglas's MD Explorer and MD520N/630N, for example, use the No Tail Rotor (NOTAR) system designed for a US Army helicopter competition. Most turboshaft engines also came from military development programmes.

International military helicopter spending is half

what it was ten years ago. So, helicopter makers will have fewer new products to offer as research and development funding trails off. Civil helicopter programmes are receiving fewer indirect military procurement subsidies. The world's civil helicopter builders are making the most of a bad situation. They are deriving new versions from existing helicopter designs, leveraging off-the-shelf technology, and searching for untapped civil markets. Kaman's K-MAX is a good example. It uses technologies and systems developed long ago, such as intermeshing rotor blades and the T53 engine. Kaman is hoping to exploit what it sees as an untapped market - heli-logging and outsize cargo transport.

In their search for untapped markets, helicopter makers often come back to the idea of using large helicopters as passenger transports. The last model to sell (in small numbers) for this role was Sikorsky's 30-seat S-61, but Sikorsky is hoping to revive the concept with its S-92 proposal, using technology from the US Army UH-60. Agusta/Westland is proposing a civil 'heliliner' version of its 30-passenger EH-101 military model. Eurocopter's 23/29-seat AS.332 is available as a passenger transport. Bell/Boeing is pushing a civil CTR-22 variant of its V-22 Osprey military tilt-rotor transport.

These tough times make rationalization seem

inevitable. But, as in the other civil aviation markets, rationalization in the helicopter market is problematic. France's Aerospatiale and Germany's DASA/MBB merged their helicopter units in 1992, creating Eurocopter. Yet this did not get rid of any production lines or products, even competing ones. And, thanks to Bell Helicopter Textron, Eurocopter cannot even claim to be the dominant player in the civil helicopter market. In addition to tapping any markets they may have overlooked, helicopter makers are lobbying the US government to not dump thousands of surplus-to-requirements military helicopters on the market. So far, the government has limited sales of these helicopters, but these OH-58s, UH-1s, and other used models represent a Sword of Damocles. If released to government agencies and police departments, these machines could collapse a lucrative new-sales market segment for years to come. Police sales alone are about 10% of the market for new civil helicopters.

This ex-military problem is unique to the US market. Other countries have far smaller military helicopter inventories. Yet all helicopter manufacturers, US and European, regard the US market as an important place to do business. The US absorbs as many new turbine helicopters as Europe - both buy about 25% of each year's production.

Helicopter makers increasingly look for sales to the Far East, which now also absorbs about 25% of new turbine helicopter production. As a result of this trend, many manufacturers are trying to penetrate Asian markets by offering helicopter workshares to Asian companies. Eurocopter's BK.117 was one of the first programmes with high Asian content, with Kawasaki a full partner. Several Bell models are built under licence in Japan. McDonnell Douglas has given Kawasaki responsibility for the transmission on its MD Explorer. Ultimately, Western helicopter makers may be faced with at least one Japanese competitor. As elsewhere, rationalization will just keep pace with the arrival of new market entrants.

All for the Best?

Overcapacity, therefore, is a problem in every civil aircraft market. The reason is simple. Aircraft are charming animals. It is difficult for an industry executive or a government planner to look at a blueprint for a new aircraft and resist funding it. Any aircraft programme holds the allure of corporate or national prestige - the international equivalent of having a shiny new sports car in your driveway. Aircraft programmes also promise jobs, even if these jobs come at a price which makes them inefficient. The aviation industry, more than steel, cars, and

Agusta's AS 61N1 Silver, developed from the Sikorsky S 61N, carries up to 28 passengers

computers, makes the people involved tremble with feelings of power and progress.

It is difficult to see what will reduce the overcapacity which results. Yet consider the alternative. A return to the booming defence aerospace markets of the 1980s (or, more ominously, the 1940s) would absorb a lot of excess civil aircraft capacity. It would also keep companies occupied with new fighters and bombers rather than transports and business jets. Therefore, perhaps it is for the best that we have so many civil aircraft. It probably keeps us out of trouble.

Aerospatiale/BAC Concorde

Concorde remains the only supersonic airliner in the World, with 14 currently in service

Most jetliners today look about the same - podded engines, thin, swept wings, and oval fuselage. However, at major airports (London, New York, Paris, etc.), you will sometimes spot (and hear) an unusual plane with an elegant delta wing, droppable nose, and very loud noise footprint. "Hey! It's the Concorde!" is the correct thing to say. Concorde is the only supersonic transport (SST) aircraft in the world. A four-engine narrowbody, it was designed and built by France's Aerospatiale and British Aircraft Corporation (BAC; now British Aerospace). SST studies began in 1955. The British and French governments agreed to

pool their efforts under an agreement signed in November 1962. The first test aircraft flew in March 1969. The Concorde entered service in January 1976.

Meanwhile, a competing US SST programme was cancelled. The USSR developed the Tupolev Tu-144 SST, now out of service. It costs a lot to fly supersonically - more than twice the cost of subsonic first-class tickets. The Anglo-French team foresaw a market for 200 Concordes; predictably, they sold 16, to Air France and British Airways. Both airlines still use most of these.

The Concorde will remain in service until around 2010. After that, several technology development projects are under way to develop a larger successor aircraft. Concorde's poor market showing provides the largest disincentive for participants in these efforts.

Specifications (Concorde):

Powerplant: four Rolls-Royce/SNECMA Olympus 593 Mk 602 turbojets, each rated at 169.3 kN (38,050 lbst)
Dimensions: length: 61.66 m (202 ft 3.6 in); height: 1.96 m (6 ft 5 in); wing span: 25.56 m (83 ft 10 in)
Weights: empty operating: 78,700 kg (173,500 lb); MTOW: 185,065 kg (408,000 lb)
Performance: cruise speed: 2,179 km/h (1,176 kts); range: 6,380 km (3,970 nm)
Passengers: 128

Airbus A300

The A300 was the first Airbus jet transport, and the first twin engine widebody airliner

A mid-sized medium-range jetliner, the A300 was the first widebody twinjet. It was also the first plane built by the Airbus consortium, and as such was the start of some serious troublemaking in the commercial jetliner market. Airbus began in 1965 as an Anglo-French project, and was finalized later as a French/German/British/Spanish effort with final production facilities in Toulouse. First proposed in 1968, the A300 made its first flight in October 1972. French and West German certification came in March 1974, and the A300 entered service, with Air France, in May.

The A300B2 and B4 were the first two variants.

Airbus built 248 of these, with production ending in late 1984. They were replaced by the current A300-600, also available as the extended-range -600R. The -600 features a two-crew flight deck, increased passenger and freight capacity, and other improvements. In addition to Air France, big users of the A300 include Lufthansa, Egyptair, American Airlines, and Thai Airways. In 1991, Federal Express launched a freighter variant, the A300-600F. This order, for at least 25 aircraft, will help keep A300 production going into the next century.

Airbus is also building four huge A300-600ST Super Transporters to move aerostructures between production facilities. These enormous beasts are called Belugas, for reasons obvious to anyone who sees them.

Specifications (A300-600R):

Powerplant: two General Electric CF6 or Pratt & Whitney PW4158 turbofans.
Data below is for aircraft with CF6-80C2A5s, each rated at 273.6 kN (61,500 lbst)
Dimensions: length: 54.08 m (177 ft 5 in); height: 16.53 m (54 ft 3 in); wing span: 44.84 m (147 ft 1 in)
Weights: empty operating: 89,813 kg (198,003 lb); MTOW: 170,500 kg (375,885 lb)
Performance: cruise speed: 875 km/h (472 kts); range: 7,410 km (4,000 nm)
Passengers: 250 (three class)

Airbus A310

The A310 is a longer-ranged version of the A300

The A310, basically a shortened A300, was Airbus's second project. It began life in the early 1970s as the A300B10 design study, and was launched in July 1978. The A300B9 and B11 studies, incidentally, became the A330 and A340 respectively.

The first A310 flew in April 1982. French/West German certification came in March 1983, and the type entered service with Swissair and Lufthansa in April. Like the A300, the A310 is available with a choice of General Electric or Pratt & Whitney engines.

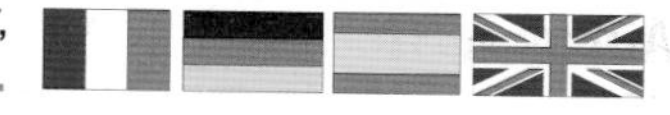

The first A310 variant was the -200, followed by the longer-range -300, which first flew in July 1985. Airbus built 85 -200s, but production of this variant is basically over. Airbus has built about 260 A310s so far. Big users include Singapore Airlines, Turkey's THY, Delta Air Lines, Air France, and Lufthansa.

In recent years, the A310 did its part to end the Cold War. In 1988, East Germany's Interflug bought three -300s, replacing hopeless Ilyushin Il-62s. These A310s are now used by the German Air Force. In 1991, the A310 became the first Western airliner granted Russian State Aviation certification. Production of the A310 is down to a few aircraft per year, but because the type is built alongside the A300, the A310 could survive past the year 2000.

Specifications (A310-300):

Powerplant: two General Electric CF6 or Pratt & Whitney PW4152 turbofans.

Data below is for aircraft with CF6-80C2A2s, each rated at 238 kN (53,500 lbst)

Dimensions: length: 46.66 m (153 ft 1 in); height: 15.8 m (51 ft 10 in); wing span: 43.89 m (144 ft)

Weights: empty operating: 80,344 kg (177,128 lb); MTOW: 150,000 kg (330,695 lb)

Performance: cruise speed: 875 km/h (472 kts); range: 7,982 km (4,310 nm)

Passengers: 210 (three class)

Airbus A320

The A320 was the first narrowbody from Airbus

The A320 was Airbus's first narrowbody jetliner. Designed to carry 150 passengers on short-to-medium routes, the A320 competes with Boeing's 737 and McDonnell Douglas's MD-80/90. Airbus was late in arriving to the narrowbody trunkliner market. The consortium had been building widebodies for almost 15 years by the time it launched the A320 programme in 1984. But, realizing that the market was ready for a new-technology trunkliner, Airbus designed the A320 with fly-by-wire controls and 15% composite materials content. Airbus also gave customers a choice of engines - General Electric/Snecma's CFM56, or International Aero Engines V2500.

The launch order came from Pan Am. Fortunately,

this did not doom the programme. The A320 made its first flight in early 1987. In March 1988 Air France and British Airways took delivery of the first two A320s. These are the only Airbuses British Airways operates, and it only has them because it acquired British Caledonian, which ordered ten.

The first version was the A320-100. Only 21 were built before production switched to the A320-200, which is distinguished by wingtip fences. Airbus also builds shortened and stretched variants of the A320, known as the A319/321.

Airbus has built over 500 A320s, and production is continuing. Major users include United Airlines, Northwest Airlines, Lufthansa, Air France, Indian Airlines, Air Inter, and Air Canada.

Specifications (A320-200):

Powerplant: two CFM International CFM56-5 or International Aero Engines V2500-A1 turbofans.
Data below is for aircraft with CFM56-5A1s, each rated at 111.2 kN (25,000 lbst).
Dimensions: length: 37.57 m (123 ft 3 in); height: 11.80 m (38 ft 8.5 in); wing span: 33.91 m (111 ft 3 in)
Weights: empty operating: 41,782 kg (92,113 lb); MTOW: 75,500 kg (166,449 lb)
Performance: cruise speed: 903 km/h (487 kts); range: 5,000 km (2,700 nm)
Passengers: 150 (two class)

Airbus A319/321

The 124-seat A319 is derived from the A320

The A319 and 321 are, respectively, the shortened and stretched versions of Airbus's A320 narrowbody jetliner. Both use the same systems, wings, and engine selection as the 150-seat A320. The A319 and 321 are built at Daimler Benz Aerospace's facility in Hamburg, while the A320 and all other Airbuses are assembled at Aerospatiale's Toulouse plant. The A321 stretch concept began at the start of the A320 programme. Airbus realized that with two fuselage plugs and some modifications, it could create a 186-seat competitor to the Boeing 757. It began marketing the idea in May 1989, and launched the

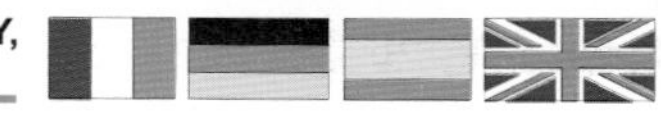

A321 in September.

The A321 made its first flight in March 1993, followed by European JAA certification in December. The first production aircraft was delivered to Lufthansa in January 1994.

By mid 1995 Airbus had delivered over 30 A321s. Customers include Alitalia, which has ordered 40, Air Inter, and ILFC.

The 124-seat A319 is a more recent effort, first conceived in the early 1990s. Airbus had only six firm orders from ILFC when it launched the A319 in June 1993, the middle of a major aircraft industry depression. Other orders have arrived from Swissair, Air Canada, and Air Inter. A319 deliveries will start in Spring 1996.

Specifications (A319):

Powerplant: two CFM International CFM56-5 or International Aero Engines V2522-A5 turbofans.
Data below is for aircraft with CFM56-5A4s, each rated at 97.9 kN (22,000 lbst)

Dimensions: length: 33.80 m (110 ft 11 in); height: 11.80 m (38 ft 8.5 in); wing span: 33.91 m (111 ft 3 in)

Weights: empty operating: 40,125 kg (88,460 lb); MTOW: 64,000 kg (141,095 lb)

Performance: cruise speed: 903 km/h (487 kts); range: 5,000 km (1,900 nm)

Passengers: 124 (two class)

Airbus A330

Dragonair's Airbus A330s are the first of the series to have Rolls Royce engines

The A330 is a twin-engine medium-range widebody jetliner built by the Airbus consortium. Closely related to the four-engine A340, the A330 holds 335 passengers in a three-class configuration. Airbus launched the A330/340 programme in June 1987. The A330 was rolled out in Toulouse in October 1992, followed by a first flight in November. Four prototypes were built. On 21 October 1993 the A330 became the first airline to obtain joint US/European FAA/JAA certification.

In late 1993 Airbus delivered the first production A330 to France's Air Inter. By mid 1995 over 25 A330s

were delivered. Major users include Cathay Pacific, which has ordered ten, Northwest Airlines, which has ordered 16, and Malaysia Airlines, which has ordered 10. The A330 is powered by the customer's choice of turbofans. Options include General Electric's CF6-80E1, Rolls-Royce's Trent 700, and Pratt & Whitney's PW4168. The first production model is the A330-300. It competes with the Boeing 767/777 and McDonnell Douglas MD-11.

For the future, Airbus is considering stretched and shortened versions of the A330. An A330 stretch could carry up to 400 passengers. A shortened version, possibly designated A329, could replace the A300 and A310 in Airbus's product line.

Specifications (A330-300):

Powerplant: two General Electric CF6-80E1A2, Pratt & Whitney PW4164/4168, or Rolls-Royce Trent 768/772 turbofans.
Data below is for aircraft with CF6-80E1A2s each rated at 300.3 kN (67,500) lbst

Dimensions: length: 63.65 m (208 ft 10 in); height: 12.92 m (42 ft 5 in); wing span: 45.23 m (148 ft 5 in)

Weights: empty operating: 120,285 kg (265,183 lb); MTOW: 212,000 kg (467,379 lb)

Performance: cruise speed: 850 km/h (459 kts); range: 8,334 km (4,500 nm)

Passengers: 335 (three class)

The A340 seats up to 295 and is used on long routes

The A340, the Airbus consortium's first four-engine design, is a long-range widebody competing with Boeing's 777 and McDonnell Douglas's MD-11 in the mini-jumbo market. It is closely related to the A330, the main difference being the propulsion philosophy (two big engines versus four medium ones). Many of A330/340 systems, and most of the fuselage and wings, are identical.

Originally known as the A300B11/TA11, the A340 programme was launched in June 1987. It made its first flight, at Aerospatiale's Toulouse plant, in

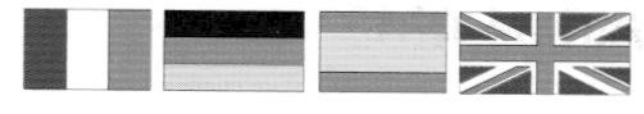

October 1991. The A340 received European JAA certification in December 1992, and entered service in March 1994. By the end of 1994, Airbus had delivered 47 A340s.

There are two basic versions of the A340. The very long-range -200 seats 262-303 passengers. The stretched -300 seats 295-335 passengers. The -300 is available as a combi aircraft, seating 195 passengers and holding six freight pallets. Ultimately, there could be a further stretch version, the -400. It could seat up to 385 passengers, but might require different engines.

Thanks to its Trans-Pacific range, the A340 has been popular with Far East Asian carriers, such as All Nippon Airways, Cathay Pacific, China Eastern, and Singapore Airlines. Other big customers include Air France, Lufthansa, and Gulf Air.

Specifications (A340-300):

Powerplant: four CFM International (General Electric/SNECMA) CFM56-2-C1 turbofans, each rated at 97.9 kN (22,000 lbst)

Dimensions: length: 57.12 m (187 ft 5 in); height: 12.92 m (42 ft 5 in); wing span: 45.23 m (148 ft 5 in)

Weights: empty operating: 75,500 kg (166,500 lb); MTOW: 161,025 kg (355,000 lb)

Performance: cruise speed: 850 km/h (459 kts); range: 8,950 km (4,830 nm)

Passengers: 295 (three class)

Boeing 707

The Boeing 707 prototype makes its lonely final flight back to Seattle

The 707 was not the first commercial jet-powered transport in the world; that honour goes to the De Havilland Comet. The 707, however, was the first truly successful effort to design an efficient, large capacity jetliner capable of crossing the Atlantic. The 707 introduced the thin, swept wing with podded engines underneath that we take for granted today.

A four-engine long-range narrowbody design, the 707 began as Boeing's Model 367-80, a prototype which flew in July 1954. The first model, the 707-120, was certified in September 1958. It entered

service with Pan Am one month later.

First 707 versions used Pratt & Whitney JT3C turbojets, followed by JT4As. In 1960 Boeing introduced JT3D turbofans as an option. These later became standard, but Boeing also built the 707-420, powered by Rolls-Royce Conways. The most popular 707 was the intercontinental -320; Boeing built 580 707-320s.

Obviously, the 707 could not go on forever. The last of 878 commercial versions was delivered to Morocco in 1982. Many Third World carriers, such as LanChile, Egyptair, and TransBrasil, still use small numbers of 707s. The 707 also found extensive use for military applications, most notably as an AWACS radar plane. The production line for these versions closed in April 1991. Final count: 1,010 707s.

Specifications (707-320C):

Powerplant: four Pratt & Whitney JT3D-7 turbofans, each rated at 84.5 kN (19,000 lbst)

Dimensions: length: 46.61 m (152 ft 11 in); height: 12.93 m (42 ft 5 in); wing span: 44.42 m (145 ft 9 in)

Weights: empty operating: 66,406 kg (146,400 lb); MTOW: 151,315 kg (333,600 lb)

Performance: cruise speed: 973 km/h (525 kts); range: 9,265 km (5,000 nm)

Passengers: 147 (two class)

Boeing 727

Boeing built 1,832 727 trunkliners and the type remains ubiquitous in the United States

The 727, a 145-seat three-engine transport for domestic trunk routes, was Boeing's second jetliner after the 707. After examining the alternatives, Boeing decided to use a rear-engine trijet configuration, an idea borrowed from the smaller Hawker Siddeley Trident. All 727s are powered by members of the Pratt & Whitney JT8D family and have a three-crew flight deck. The 727 programme was launched in December 1960 with orders from Eastern and United Airlines. A prototype flew in February 1963. Revenue service began in February 1964.

The first 727 was the -100, a 131-seat model powered by JT8D-1s. The -100 was also available as the -100C, a convertible cargo/passenger model, and the -100QC, a Quick Change cargo version. The 145-seat stretched -200 was certified in November 1967. It became the standard 727. The last variant was the pure-freight 727F, delivered in 1983.

Boeing built a total of 1,832 727s, including one test aircraft. Of these, 1,245 were 727-200s and 15 were 727Fs. The last 727 was delivered in September 1984.

Many 727s are still in service, with American, Continental, Delta, TWA, Northwest, United, and many others. Federal Express and UPS use the cargo versions. A number of re-engining and hush-kitting options are available to keep the type in service past 2000.

Specifications (727-200):

Powerplant: three Pratt & Whitney JT8D-9A turbofans, each flat rated at 64.5 kN (14,500 lbst); also available with the uprated JT8D-11, -15, -17, and -17R.

Dimensions: length: 46.69 m (153 ft 2 in); height: 10.36 m (34 ft); wing span: 32.92 m (108 ft)

Weights: empty operating: 45,360 kg (100,000 lb); MTOW: 83,820 kg (184,800 lb)

Performance: cruise speed: 872 km/h (471 kts); range: 3,706 km (2,000 nm)

Passengers: 145 (two class), 189 (all economy)

Boeing 737-100/200

The first 737 series, a 100/120 seat design, used the Pratt & Whitney JT8D engine

The legendary 737 series began with the 737-100 version, a 'Baby Boeing' carrying 100 passengers on short routes. Boeing began the 737-100 programme in November 1964, even though the similar BAC 1-11 and Douglas DC-9 programmes were well under way. Like the DC-9, the early 737s were powered by Pratt & Whitney JT8D engines.

In February 1965 Lufthansa placed a launch order for 21 737-100s. A prototype flew in April 1967. Meanwhile, Boeing decided to stretch the 737, creating the 120/130-seat 737-200. United launched

this version in April 1965, and a 737-200 flew in August 1967. The 737-100 and -200 were certified in December 1967. The -100 entered service in February 1968, followed by the -200 in April 1968.

The programme's highs and lows tell the story of a very cyclical industry - Boeing built 114 737s in 1969, and only 22 in 1972. By 1981 annual production rose to 108 aircraft. Boeing delivered the last 737-200 in 1988. By then, 737-300 production was well under way. A total of 1,114 -200s were built, plus 30 -100s. Over 100 of these were 737-200C convertible cargo versions. Many 737-200s are still in service, from Algeria to Zambia. Big users include the US majors, especially Delta, United, and USAir. Several hush-kit programmes are available to allow the series to meet Stage 3 noise restrictions.

Specifications (B737-200):

Powerplant: three Pratt & Whitney JT8D-9A turbofans, each rated at 64.5 kN (14,500 lbst)
Dimensions: length: 46.60 m (153 ft 2 in); height: 10.36 m (34 ft 0 in); wing span: 32.92 m (108 ft 0 in)
Weights: empty operating: 45,360 kg (100,000 lb); MTOW: 83,820 kg (184,800 lb)
Performance: cruise speed: 856 km/h (462 kts); range: 3,437 km (1,855 nm)
Passengers: 120-130 (two class)

Boeing 737-300/400/500

The first variant of the 737 series is a 128-seat model

Boeing's second generation of 737s was launched in March 1981. The second series features all-new engines, wing modifications, and a new flight deck. It competes with Airbus's A320 and McDonnell Douglas's MD-80 series. The first new 737 was the -300, launched in March 1981. Stretched to seat 128 passengers in two classes, the -300 flew in February 1984 and was certified in November 1984. Two years later, Boeing launched the 737-400, a 146-seat stretch. It entered service in September 1988. Finally,

Boeing launched the shortened 108-seat 737-500 in May 1987. It entered service in March 1990.

The second 737 series has proven even more popular than the 737-100/200. As of late 1995 Boeing had delivered over 1,600 737-300/400/500s. Many of the world's airlines use the type. Big users include Delta, Continental, Lufthansa, MAS Malaysia, United, and USAir.

US discount carrier Southwest has made its name flying only 737s, with orders for over 165 -300s and -500s. Upon spotting a 737 for a different airline, a Southwest representative at the Boeing plant reportedly asked: "What's that plane doing at our factory?" Production of all three current 737s is continuing, but slowing. The replacement 737-600/700/800 will be here in a few years.

Specifications (737-300):

Powerplant: two CFM International CFM56-3C-1 turbofans rated at 88.97 kN (20,000 lbst)
Dimensions: length: 33.40 m (109 ft 7 in); height: 11.13 m (36 ft 6 in); wing span: 28.88 m (94 ft 9 in)
Weights: empty operating: 31,895 kg (70,320 lb); MTOW: 56,472 kg (124,500 lb)
Performance: cruise speed: 794 km/h (429 kts); range: 4,554 km (2,830 nm)
Passengers: 128 (two class)

Boeing 737-600/700/800

Like the second 737 series, the third series is manufactured in three discrete variants

Boeing is now working on a third 737 generation designed to replace the 737-300/400/500. Specifically, the 108-seat 737-600 will replace the -500, the 146-seat -700 will replace the -300, and the 160-seat -800 will replace the -400. Boeing announced the 737-X series in June 1993. In November 1993 Southwest Airlines launched the 737-700 with 63 firm and 63 option orders. The next to be launched was the -800. Hapag Lloyd ordered 16 of these in September 1994. Finally, SAS launched

the -600 in March 1995, with 35 firm and 35 option orders. Each of these launches was accompanied by a Boeing announcement of a 737-300/400/500 production rate cut.

Additional 737-700/800 orders have come from Maersk, Air Berlin, Bavaria, and Germania. Boeing hopes to begin 737-700 deliveries in October 1997.

The new 737 series will feature quieter and more efficient CFM56-7 versions of the CFM56-3 engines used on the current 737s. They should permit the new 737s to have operating costs 15% lower than the current series, and to meet Stage 4 noise regulations. The next 737s will also have bigger wings, new avionics, and more flexible interiors.

Thanks to these efforts, the 737 family will probably still be in production in 2015, its 50th year.

Specifications (737-700):

Powerplant: two CFM International CFM56-7 turbofans each rated at 106.76 kN (24,000 lbst)

Dimensions: length: 33.63 m (110 ft 4 in); height: 11.13 m (36 ft 6 in); wing span: 34.31 m (112 ft 7 in)

Weights: empty operating: 31,895 kg (70,320 lb); MTOW: 67,585 kg (149,000 lb)

Performance: cruising speed: Mach 0.80; range: 5,556 km (3,000 nm)

Passengers: 128 (two class)

Boeing 747

The 747 has revolutionized the airline industry

'Building A Legend' says the banner above Boeing's 747 assembly line. The banner is right. The 747 is the largest commercial jetliner in the world. It was also the first widebody airliner, making it possible for vast numbers of people to travel cheaply to distant parts of the world. It has brought us closer together. The 747 story began in the 1960s, when Boeing lost a US Air Force competition to build a large transport. The company offered its design as a civil jetliner. Pan Am launched the programme in April 1966. The 747 was certified in December 1969.

The first model was the 747-100, powered by Pratt & Whitney JT9D turbofans. It was followed by the longer range -200, which entered service in 1971. The -300 featured a stretched upper deck. It entered service in 1983. The current model is the 747-400, distinguished by winglets. It also has greater range, wider wings, and an advanced two-crew flight deck. It is available in cargo and combi variants. Boeing is considering several new stretched and re-winged variants for the future.

As of late 1995, Boeing had built over 1,070 747s, including over 340 -400s. Almost every major international airline uses 747s. Big users include British Airways, Japan Airlines, Singapore Air, and Korean Airlines.

Specifications (747-400):

Powerplant: four General Electric CF6-80C2, Pratt & Whitney PW4000 or Rolls-Royce RB.211-524 turbofans.
Data below is for aircraft with PW4056s rated at 258 kN (57,900 lbst).

Dimensions: length: 70.66 m (231 ft 10 in); height: 19.41 m (63 ft 8 in) wing span: 64.92 m (213 ft)

Weights: empty operating: 180,985 kg (399,000 lb); MTOW: 391,500 kg (870,000 lb)

Performance: cruise speed: 940 km/h (507 kts); range: 13,278 km (7,165 nm)

Passengers: 421 (three class)

Boeing 757

American, Delta, United and other US carriers are the biggest users of the Boeing 757

The 757 is a medium-range twinjet airliner designed for transcontinental operations. The largest narrowbody built, the 757 can carry up to 220 passengers. Boeing created the 757 as a successor to the 727. It launched the programme in August 1978, with orders from British Airways and Eastern Air Lines. The 757 was rolled out in January 1982, and made its first flight one month later. It entered service with Eastern in January 1983.

The first version was the 757-200. It remains the current production model, and is available as a cargo

aircraft. Originally, Boeing planned to offer a shortened 150-seat 757-100. A stretched 757-300 could arrive in the late 1990s, but this has yet to be confirmed.

The 757 fuselage cross section is basically the same used on Boeing's 707, 727, and 737, leading some to call Boeing's Renton facility 'the Great Fuselage Machine'. The 757 can be powered by a choice of Rolls-Royce or Pratt & Whitney engines, and was one of the first planes to use a two-crew flight deck.

The same flight deck can be found on the Boeing 767, which was designed along with the 757. Boeing has built over 680 757s, and production is continuing. Major 757 users include American, Delta, Northwest, and United Parcel Service.

Specifications (757-200):

Powerplant: two Rolls-Royce RB.211-535E4 turbofans each rated at 178.4 kN (40,100 lbst)

Dimensions: length: 47.32m (155 ft 3 in); height: 13.56 m (44 ft 6 in); wing span: 38.05 m (124 ft 10 in)

Weights: empty operating: 57,180 kg (126,060 lb); MTOW: 99,790 kg (220,000 lb)

Performance: cruise speed: 851 km/h (459 kts); range: 5,222 km (2,820 nm)

Passengers: 186 (two class)

Boeing 767

The stretched, 260-seat Boeing 767-300 has eclipsed the original 767-200

The 767 is Boeing's smallest widebody, a 220-270 passenger twinjet which competes with (and resembles) Airbus's A310. The launch order came from United Airlines in July 1978, and the 767 first flew in September 1981. First deliveries came in August 1982.

The first 767 was the -200, but a stretched 767-300 was launched in September 1983. The -300 is 21 ft (6.4 m) longer, but is otherwise similar to the -200. Most production today is for the 767-300. Both

models are available as extended range variants, as the -200ER/-300ER. ER variants include larger wing centre-section fuel tanks, and structural changes needed to support the extra weight.

The 767 is available with a choice of General Electric, Pratt & Whitney, or Rolls-Royce turbofans. Most of the fuselage is built in Japan by Kawasaki and Mitsubishi. The 767 was the first Boeing plane to use a two-crew flight deck with electronic flight instrument systems (EFIS). Seating is usually seven or eight abreast. Over 575 767s were built by mid 1995. Major users include United, All Nippon Airways, American Airlines, and British Airways. UPS has launched a freighter version, the -300F.

Specifications (767-300):

Powerplant: Two General Electric CF6-80C2, Pratt & Whitney PW4050/4052, or Rolls-Royce RB.211-524G turbofans.
Data below is for 767-300s with CF6-80C2B2s, each rated at 233.5 kN (52,500 lbst).
Dimensions: length: 54.9 m (180 ft 3 in); height: 15.9 m (52 ft); wing span: 47.6 m (156 ft 1 in)
Weights: empty operating: 86,953 kg (191,700 lb); MTOW: 156,490 kg (345,000 lb)
Performance: cruise speed: 850 km/h (459 kts); range: 7,450 km (4,020 nm)
Passengers: 260 (210 on -300ER)

Boeing 777

Cathay Pacific's first Rolls Royce-powered Boeing 777 seen here touring Seattle

The last new large jetliner entering service this century, the Triple Seven is a widebody twinjet seating 300-350 passengers, the 777 is designed for intercontinental and transcontinental routes. It fills the gap in Boeing's product line between the 767 and 747.

Boeing announced the 767-X project in June 1989. The new plane was redesignated 777 after its launch in October 1990. The first 777 flew in June 1994. It was certified in April 1995. The 777-200A entered service with launch customer United Airlines in June

1995. It will be followed by the longer range -200B in late 1996. Other major customers include British Airways, All Nippon Airways and Saudia. As with the 767, the 777 is available with a choice of General Electric, Pratt & Whitney, or Rolls-Royce turbofans. It is the first application for GE's GE90 engine. Japanese industry also has a considerable stake in the 777 project, building most of the 777's fuselage. Fun fact: the 777's engines, the most powerful aero engines ever built, are housed in nacelles as wide as a 737's fuselage.

The 777-300 stretch variant, due to enter service in 1998, will seat up to 440 passengers and will require engines rated up to 436 kN (98,000 lbst). It has been ordered by Thai Airways, All Nippon, and Cathay Pacific.

Specifications (777-200A):

Powerplant: two General Electric GE90, Pratt & Whitney PW4000 or Rolls-Royce Trent 800 turbofans.

Data below is for aircraft with PW4074s rated at 329.17 kN (74,000 lbst).

Dimensions: length: 63.73 m (209 ft 1 in); height: 18.51 m (60 ft 9 in); wing span: 60.93 m (199 ft 11 in)

Weights: empty operating: 135,580 kg (298,900 lb); MTOW: 229,520 kg (506,000 lb)

Performance: cruise speed: 897 km/h (484 kts); range: 7,505 km (4,050 nm)

Passengers: 305-328 (three class)

British Aerospace 1-11

The 1-11 is still flown, McDonnell Douglas uses this 1-11 as an avionics testbed

The One-Eleven is a 65-119 seat narrowbody twinjet airliner designed by British Aircraft Corporation (now British Aerospace). The 1-11 was BAC's answer to Douglas's DC-9 and Fokker's 28. The 1-11 began in the 1950s with Hunting Aviation's proposed H.107 48-seat jetliner. When BAC acquired Hunting in 1960, it enlarged the design, allowing seating for 65 passengers. A prototype flew in August 1963. The first 1-11 was the Series 200, which was certified in April 1965. It was followed by the Series 300 and

400. The largest model was the Series 500, a stretched 300/400 with wider wings and more powerful engines. It carried up to 119 passengers and entered service in November 1968.

UK production of the 1-11 ended in the late 1970s, with a total of 230 aircraft built. In 1982, British Aerospace helped CNIAR (now Romaero) establish a 1-11 assembly line in Romania. About 20 were built there, the first flying in September 1986. Romaero has developed a new version, the Airstar 2500, with Rolls-Royce Tay engines. If this proceeds it could enter service in 1997 or 1998.

As of 1995, over 100 1-11s remain in service. British World has about 10, European Aviation Air Charter has about 20, and Maersk UK has six. BAe continues to support the type with spares.

Specifications (1-11-500):

Powerplant: two Rolls-Royce Spey Mk 512 DW turbofans, each rated at 54.5 kN (12,550 lbst)
Dimensions: length: 32.61 m (107 ft 0 in); height: 7.47 m (24 ft 6 in); wing span: 28.5 m (93 ft 6 in)
Weights: empty operating: 24,758 kg (54,582 lb); MTOW: 45,200 kg (99,650 lb)
Performance: cruise speed: 742 km/h (400 kts); range: 3,484 km (1,880 nm)
Passengers: 119

The 146/Avro production line at Woodford

The 146/RJ is a four-engine 70-115 seat jetliner designed for short-haul operations. It is instantly identifiable as the only civil jetliner with a high wing (mounted above the fuselage). The 146/RJ has a complex history and nomenclature. It was designed as the Hawker Siddeley HS 146, but this project collapsed. When Hawker became part of BAe, it became the BAe 146. BAe relaunched the 146 in July 1978. It flew in September 1981 and entered service in May 1983.

In June 1992 BAe designated each aircraft in the

series RJ (regional jet) followed by the seating capacity. The RJ70 and RJ85 uses the 146-100 airframe, while the RJ100 uses the 146-200 and the RJ115 uses the 146-300. BAe also created a separate division, Avro, to market the RJ series. RJs are also called Avroliners.

The new RJs feature improved engines, new interiors, and digital avionics. The first RJ flew in March 1992. Deliveries began in late 1993. BAe built about 220 146s. Over 100 of these are on lease to various operators, but TNT Express owns 18, Air Wisconsin has 12, and Air UK has 12. As of late 1995 BAe had built about 40 RJs. RJ orders have come from Lufthansa Cityline, Sabena, and Crossair. RJ production is continuing.

Specifications (Avro RJ100):

Powerplant: four AlliedSignal LF507 turbofans, each rated at 31.14 kN (7,000 lbst)
Dimensions: length: 28.6 m (93 ft 10 in); height: 8.61 m (28 ft 3 in); wing span: 26.21 m (86 ft 0 in)
Weights: empty operating: 24,993 kg (55,100 lb); MTOW: 46,039 kg (101,500 lb)
Performance: cruise speed: 711 km/h (384 kts); range: 2,593 km/h (1,400 nm)
Passengers: 100

Canadair Regional Jet

The RJ is a popular alternative to turboprops

Canadair's Regional Jet (RJ) is the first of its kind: a 50-seat transport designed for long, thin regional routes. It is derived from the company's Challenger business jet, and uses the same General Electric CF34 turbofans. Seating is four abreast.

The RJ programme was launched in March 1989, and the aircraft first flew in May 1991. Canadian certification was awarded in July 1992, followed by US FAA and European JAA certification in January 1993.

Major users include Comair, Lufthansa, Skywest,

and Air Canada. The RJ is replacing both turboprop transports and aging small jets, such as the DC-9 and Fokker 28. The RJ costs about $17 million. The baseline RJ100 has a range of 1,816 km (980 nm), while the RJ100ER extended range variant can fly 3,002 km (1,620 nm). Canadair is working on a stretch version of the RJ. The CRJ-X will seat 70 passengers, and use improved versions of the CF34.

There are currently in use no direct competitors to the RJ. However, Embraer's EMB-145 will soon arrive, and will provide some competition. Short Brother's FJX was to attack the same market, but this proposal was quashed when Bombardier, Canadair's parent company, purchased Shorts. As a consolation prize, Shorts was given the contract to build much of the RJ fuselage.

Specifications (RJ100):

Powerplant: two General Electric CF34-3A1 turbofans, each rated at 41.01 kN (9,220 lbst) with APR (automatic power reserve)

Dimensions: length: 26.77 m (87 ft 10 in); height: 6.22 m (20 ft 5 in); wing span: 21.21 m (69 ft 7 in)

Weights: empty operating: 13,653 kg (30,100 lb); MTOW: 21,523 kg (47,450 lb)

Performance: cruise speed: 786 km/h (424 kts); range: 1,816 km (980 nm)

Passengers: 50

Embraer EMB-145

The EMB-145 will compete with Canadair's RJ

The EMB-145 Amazon is a 48-seat twinjet regional aircraft built by Brazil's Embraer. The new aircraft is a stretched EMB-120 with new wings and engines. Embraer unveiled the EMB-145 Amazon in June 1989 at the Paris Air Show. The original EMB-145 design used the EMB-120's straight wings with turbofans mounted above the wings. In March 1991, Embraer changed the design to include new swept wings with engines mounted beneath the wings. In December 1991 Embraer approved a second redesign

of the EMB-145, with engines on the rear fuselage.

Spain's GAMESA is building the wings. Allison engines is providing its AE 3007 turbofan. There are 72 other risk-sharing partner/suppliers, so Embraer is basically just the EMB-145 final integrator.

In June 1993, Embraer gave the go-ahead to the Amazon project. A prototype was rolled out and flown in August 1995, over two years later than originally planned. Embraer will deliver the first EMB-145s in December 1996. Embraer has committed itself to building 24 Amazons in 1997, but as of late 1995 only about 18 orders and 19 options had been received. The firm orders include Australia's Flight West (4), TransBrasil (4), and France's Regional (3). Embraer also claims to have letters of intent for 127 more Amazons.

Specifications (EMB-145):

Powerplant: two Allison AE 3007A turbofans, each rated at 31.32 kN (7,040 lbst)

Dimensions: length: 27.93 m (91 ft 7.5 in); height: 6.71 m (22 ft 0.25 in); wing span: 20.04 m (65 ft 9 in)

Weights: empty operating: 11,585 kg (25,540 lb); MTOW: 19,200 kg (42,329 lb)

Performance: cruise speed: 760 km/h (410 kts); range: 1,482 km (800 nm)

Passengers: 50 (one class)

Fokker 28

Over 150 F28s are still flying today

The F28 Friendship is a twin engine narrowbody jetliner built by Fokker of the Netherlands. A 65-85 seat short/medium-range design, the F28 competed with Douglas's DC-9 and BAC's 1-11. Like the 1-11, the F28 uses Rolls-Royce Spey engines mounted on the rear fuselage.

The F28 programme began in the early 1960s, with the first prototype flying in May 1967. Germany's LTU was the launch customer, and the airline received its first F28 just after certification in February 1969.

The first F28 was the Mk 1000, a 65-seat model. It

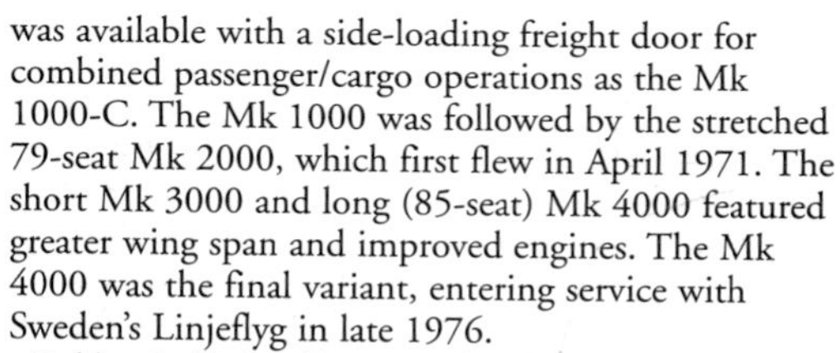

was available with a side-loading freight door for combined passenger/cargo operations as the Mk 1000-C. The Mk 1000 was followed by the stretched 79-seat Mk 2000, which first flew in April 1971. The short Mk 3000 and long (85-seat) Mk 4000 featured greater wing span and improved engines. The Mk 4000 was the final variant, entering service with Sweden's Linjeflyg in late 1976.

Fokker built 241 F28s. Production ended in late 1986, but Fokker went on to build the derivative F100. Over 150 F28s are still in service, with USAir, SAS, Merpati, TAT, and Air Niugini.

Specifications (F28 Mk 4000):

Powerplant: two Rolls-Royce RB183-2 Mk 555-15P turbofans, each rated at 44 kN (9,900 lbst)
Dimensions: length: 26.76 m (87 ft 9.5 in); height: 8.47 m (27 ft 9.5 in); wing span: 25.07 m (82 ft 3 in)
Weights: empty operating: 17,645 kg (38,900 lb); MTOW: 33,110 kg (73,000 lb)
Performance: cruise speed: 678 km/h (366 kts); range: 2,085 km (1,125 nm)
Passengers: 85

Fokker 100/70

American Airlines operate 75 Fokker F100s

The F100 is a 107-seat stretched follow-on to the Fokker 28. The F100 features new Tay 650 engines and a digital 'glass' cockpit. F100 development began in 1983. Swissair provided a launch order in July 1984. The first F100 flew in November 1986, and Swissair received the first production F100 in February 1988. The F100 received a boost in March 1989 when American Airlines ordered 75 aircraft, the largest order in Fokker's history.

While Fokker is synonymous with the Netherlands,

the F100 is an international product. In addition to the Rolls-Royce engines, the wings come from Short Brothers in Northern Ireland, and the fuselage is built by Daimler Benz Aerospace, which is now Fokker's majority shareholder.

In 1992, Fokker decided to revive the old F28 fuselage, with the F100's systems and Tay 620 engines. This became the Fokker 70, a 70-78 seat regional jet. Fokker launched the F70 in June 1993, and deliveries began in late 1994. Fokker has built over 250 F100s as of late 1995.

Major users other than American Airlines include USAir (40), and Korean Air (12). Fokker has also built over 30 F70s, with orders from Alitalia, Air Littoral, and Sempati Air. Production of both types is continuing.

Specifications (Fokker 100):

Powerplant: two Rolls-Royce Tay Mk 620 turbofans, each rated at 61.6 kN (13,850 lbst)

Dimensions: length: 35.53 m (116 ft 6.75 in); height: 8.51 m (27 ft 10.5 in); wing span: 28.08 m (92 ft 1.5 in)

Weights: empty operating: 24,593 kg (54,217 lb); MTOW: 43,090 kg (95,000 lb)

Performance: maximum speed: 856 km/h (462 kts); range: 2,389 km (1,290 nm)

Ilyushin Il-86

The Il-86 was the first Soviet widebody transport aircraft, but is handicapped by elderly engines

The Il-86 is a four-engine commercial jet transport designed by the Ilyushin Design Bureau. A typical podded-engine design, it resembles the Airbus A340. One unique feature is that passengers enter at ground level and walk up stairs mounted inside the fuselage. This entry area also houses coat closets and baggage store containers. Il-86 design work began in the early 1970s. The first of two Il-86 prototypes flew in December 1976, and the type entered service with Aeroflot in December 1980. Its first route was

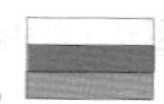

Moscow-Tashkent. In Soviet-style single-class seating configuration, the Il-86 carries 350 passengers.

While the Il-86 was the first widebody built in the Soviet Union, it was not particularly successful. Its NK-86 turbofans were antique by world standards, and the plane never met its range targets. A total of 100 Il-86s were built at the Voronezh assembly line. Many of these are inoperable. The type is only used by Russia's fragmented post-Aeroflot airlines, such as Vnukovo and Sibavia.

Due to delays with the Il-96, the Il-86 will stay in service for at least another ten years. General Electric and Snecma are marketing a re-engining programme using CFM56 turbofans. If they are successful, the Il-86 fleet could be given a second chance, with greater range and a Stage 3 noise rating.

Specifications (Il-86):

Powerplant: four KKBM Kuznetsov NK-86 turbofans, each rated at 127.5 kN (28,660 lbst)

Dimensions: length: 59.54 m (195 ft 4 in); height: 15.81 m (51 ft 10.5 in); wing span: 48.06 m (157 ft 8.25 in)

Weights: empty operating: 121,000 kg (266,200 lb); MTOW: 206,000 kg (454,150 lb)

Performance: cruise speed: 900 km/h (485 kts); range: 3,600 km (1,944 nm)

Passengers: 350 (one class)

Ilyushin Il-96

The Il-96 is now available with western avionics

The Il-96 is a four-engine widebody commercial jet transport. Designed as a follow-on to the Il-86, the Il-96 can be distinguished from its predecessor by larger engines and winglets. The Il-96 is slightly shorter and seats 300. Like the Il-86, the Il-96 resembles the Airbus A340. Il-96 design work began in the mid 1980s. The first of five prototypes flew in September 1988. The baseline version, the Il-96-300, began service in late 1992.

Due to financial problems, only a handful of Il-96s have been built at the Voronezh production line. Il-

96-300s have been delivered to Aeroflot Russian International Airlines (ARIA), and Domodedovo Air Lines.

While the Il-96-300's Perm PS-90 engines are a major improvement over the Il-86's NK-86s, Ilyushin is cooperating with American manufacturers to 'Westernize' the plane. The Il-96M variant uses Pratt & Whitney PW2337 turbofans and Rockwell-Collins avionics. It has greater range than the Il-96-300, and features a two-crew flight deck. There is also a freighter variant, the Il-96MT.

The first Il-96M, a modified Il-96-300, flew in April 1993. Il-96M/MT orders have arrived from ARIA and Partnairs, a Dutch leasing company. These orders depend on Western certification, which has been delayed for financial reasons.

Specifications (Ilyushin Il-96-300):

Powerplant: four Perm/Soloviev PS-90A turbofans, each rated at 156.9 kN (35,275 lbst)

Dimensions: length: 55.35 m (181 ft 7.25 in); height: 17.57 m (57 ft 7.75 in); wing span: 57.66 m (189 ft 2 in)

Weights: empty operating: 117,000 kg (257,940 lb); MTOW: 216,000 kg (476,200 lb)

Performance: cruise speed: 850 km/h (459 kts); range: 7,500 km (4,050 nm)

Passengers: 300 (one class)

Lockheed L-1011 TriStar

Lockheed built 250 Tristars after a troubled introduction

The L-1011 TriStar was Lockheed's only jetliner programme. A three-engine medium-capacity, medium-range widebody, it unfortunately arrived on the market at the same time as its direct competitor, Douglas's DC-10. Worse, the TriStar's development was incredibly painful. It resulted in engine manufacturer Rolls-Royce going into bankruptcy due to problems with its all-new high-bypass RB.211 turbofan. Lockheed itself was reduced to begging for a US government loan.

The company pressed on. The first L-1011 flew in November 1970, nearly five years after Lockheed began design work. The first model, the L-1011-1,

62, and 63, long-range, high-capacity versions which all entered service in 1967. Douglas built 556 DC-8s by the time production ended in May 1972.

Most DC-8s were built with Pratt & Whitney JT3D engines, as on the 707. Early DC-8s had the Pratt JT4A or JT3C, while the Series 40 offered the Rolls-Royce Conway. In the early 1980s 110 Super 61s, 62s and 63s were re-engined with CFM56 engines, becoming Super 71s, 72s, and 73s.

Most Super 70 series DC-8s are still in service, along with 50-70 other DC-8s. They are mostly used for freight operations. Big users include UPS and Airborne Express. As one aeroplane value expert commented, 25th century archaeologists will find a Super 73 flying a cargo route in South America.

Specifications (DC-8 Series 73):

Powerplant: four CFM International (General Electric/SNECMA) CFM56-2-C1 turbofans, each rated at 97.9 kN (22,000 lbst)

Dimensions: length: 57.12 m (187 ft 5 in); height: 12.92 m (42 ft 5 in); wing span: 45.23 m (148 ft 5 in)

Weights: empty operating: 75,500 kg (166,500 lb); MTOW: 161,025 kg (355,000 lb)

Performance: cruise speed: 850 km/h (459 kts); range: 8,950 km (4,830 nm)

Payload: 29,257 kg (64,500 lb) or 259 passengers

McDonnell Douglas DC-9

The DC-9 was the first US airliner to feature rear-mounted engines

The DC-9 programme began in the 1950s, as a Douglas proposal for a short/medium-range 75-seat narrowbody jetliner to complement the long-range DC-8. Douglas got the idea for a twinjet with rear fuselage-mounted engines from France's Sud-Est (now Aerospatiale), whose Caravelle jetliner was the first to use this design. Douglas launched the DC-9 programme in April 1963.

The first DC-9 was the Series 10, a 90-seat design which received FAA certification in November 1965

and entered service with launch customer Delta Air Lines in December. This was followed by the extended-wing Series 20, the 119-seat Series 30, and the 125-seat Series 40. There were also freight and convertible freight/passenger variants, and a military cargo variant, which the US Air Force designated the C-9. The final version of the DC-9 was the stretched 139-seat Series 50, which entered service in August 1975. DC-9 production ended in the early 1980s, but Douglas began building DC-9 Super 80, or MD-80.

A total of 976 DC-9s were built, including 43 C-9s. The majority of these remain in service with most major US airlines and quite a few others. The type remains popular and several airlines, including Northwest and Valujet, are hush-kitting DC-9 Series 30s for compliance with Stage 3 noise regulations.

Specifications (DC-9 Series 30):

Powerplant: two Pratt & Whitney JT8D-9s, each rated at 64.5 kN (14,500 lbst); also available with uprated JT8D-11, -15, -17.
Dimensions: length: 36.37 m (119 ft 4 in); height: 8.38 m (27 ft 6 in); wing span: 28.47 m (93 ft 5 in)
Weights: empty operating: 25,940 kg (57,190 lb); MTOW: 54,885 kg (121,000 lb)
Performance: cruise speed: 907 km/h (490 kts); range: 3,095 km (1,670 nm)
Passengers: 105-115

McDonnell Douglas DC-10

Northwest (formerly Northwest Orient) uses the DC-10 widebody on Far Eastern routes

The Douglas DC-10 is a three-engined medium/long-range widebody jetliner with seating for 255-380 passengers. Designed in the 1960s, the DC-10 was launched by orders from American and United Airlines in 1968. It made its first flight on 29 August 1970. FAA certification came in July 1971, and the DC-10 entered service with American in August.

The first model was the DC-10 Series 10, designed for US domestic service and powered by General Electric CF6 engines. The CF6-powered Series 30 was the first intercontinental version and was also available as the Series 30ER (Extended Range). The

Series 30F was a freighter variant, and the Series 10CF and 30CF were convertible freighter variants. The Series 40, entering service in late 1972, used Pratt & Whitney JT9D engines.

The DC-10 provided competition for Lockheed's L-1011. The two types carved up a market which had room for only one profitable programme. But the DC-10 did better than its nemesis, thanks in part to US Air Force procurement of KC-10A tanker/cargo transports. The DC-10 programme ended in 1989. Production totalled 446 aircraft. Of these, 60 were KC-10As. Numerous carriers still use the type, including United, American, Northwest, Japan Air Lines, and Varig. Federal Express uses DC-10 Freighters. After the DC-10, Douglas turned its attention to the derivative MD-11.

Specifications (DC-10 Series 30):

Powerplant: three General Electric CF6-50C turbofans, each rated at 227 kN (51,000 lbst)

Dimensions: length: 55.5 m (182 ft 1 in); height: 17.7 m (58 ft 1 in); wing span: 50.4 m (165 ft 5 in)

Weights: empty operating: 121,198 kg (267,197 lb); MTOW: 259,450 kg (572,000 lb)

Performance: cruise speed: 880 km/h (475 kts); range: 7,413 km (4,000 nm)

Passengers: 255-270 (mixed class), 380 (all-economy)

McDonnell Douglas MD-11

The MD-11 entered service in 1990

The McDonnell Douglas MD-11 is a three-engined long-range widebody jetliner derived from the DC-10. Design work began on a mere stretch of the older design, known as the DC-10 Series 50/60. As the project grew more ambitious, it was designated the MD-100, and in 1984, the MD-11. FAA certification was awarded in November 1990. First deliveries, to Finnair, began in December 1990. Compared with the DC-10, the MD-11 has a 5.66 m (18 ft 7 in) fuselage stretch, larger wings, more power engines, and a two-crew cockpit. As with the DC-10, the MD-11 is offered with a choice of General Electric or

Pratt & Whitney engines. Plans to offer a version with Rolls-Royce Trents fell through when MD-11/Trent launch customer Air Europe went bankrupt.

As with the DC-10, the MD-11 is available in freighter, convertible freighter, and combi variants. Federal Express operated 13 MD-11F freighter variants, and is buying more American Airlines MD-11s for freighter conversions. Korean Air Lines is also converting its five MD-11s to freighters.

McDonnell Douglas has delivered over 130 MD-11s. Major users include, Delta, Garuda, Japan Air Lines, Swissair, and Alitalia. As this is written, the future of MD-11 production is in doubt, and the production line could close as early as 1996.

Specifications (MD-11):

Powerplant: three Pratt & Whitney PW4460 or General Electric CF6-80C2 turbofans.

Data below is for aircraft with PW4460s, each rated at 266.9 kN (60,000 lbst)

Dimensions: length: 61.24 m (200 ft 11 in); height: 17.60 m (57 ft 9 in); wing span: 51.77 m (169 ft 10 in)

Weights: empty operating: 131,035 kg (288,880 lb); MTOW: 283,725 kg (625,500 lb)

Performance: cruise speed: 898 km/h (560 kts); range: 12,607 km (6,803 nm)

Passengers: 293 (three class)

McDonnell Douglas MD-80

With 1000 built, the MD-80 has surpassed the DC-9

The MD-80, a DC-9 derivative, was originally known as the DC-9 Super 80. Major changes from the DC-9 include new refanned Pratt & Whitney JT8D-200 series engines, a longer fuselage, and an increased wing span (28% greater than the DC-9-50).

Renamed to reflect Douglas's status as part of McDonnell Douglas, the first MD-80 flew in October 1979. FAA certification was granted in August 1980, and the first production aircraft went to Swissair in September.

There are five members of the MD-80 family: The

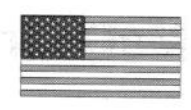

MD-81, -82, -83, and -88 all seat 155 passengers. The MD-82 has increased power engines for hot-and-high operations and the MD-83 has greater range. The MD-87 is a shortened variant seating 105-130. The MD-88 features an advanced 'glass' cockpit with EFIS displays. As with the DC-9, most US majors and many other airlines use MD-80s. These include American (over 225), Continental (60), and Alitalia (over 45). Most MD-88 production (120) went to Delta.

In June 1992, Douglas delivered its 1,024th MD-80, for a total of 2,000 twinjets including the DC-9. China's SAIC (Shanghai Aviation Industrial Corporation) built 35 MD-82s and -83s under licence. The last of these was completed in August 1994.

As this is written, production of the MD-80 is winding down. It is being replaced by the MD-90.

Specifications (MD-82):

Powerplant: two Pratt & Whitney JT8D-217 turbofans, each rated at 89 kN (20,000 lbst)

Dimensions: length: 45.06 m (147 ft 10 in); height: 9.19 m (30 ft 2 in); wing span: 32.87 m (107 ft 10 in)

Weights: empty operating: 35,369 kg (77,976 lb); MTOW: 67,812 kg (149,500 lb)

Performance: cruise speed: 813 km/h (439 kts); range: 4,032 km (2,176 nm)

Passengers: 155 (two class)

McDonnell Douglas MD-90

Delta remains the biggest customer for the MD-90

The MD-90, the third major incarnation of Douglas's twinjet family, was originally an innovative, high-tech design. It was to feature propfans, energy-efficient, ultra-high bypass jets with menacing external, swept fan blades. The company began plans for the plane in 1986. Alas, airlines proved conservative, and oil prices stayed low. Boeing also abandoned its plans for a propfan liner, the 7J7. Douglas decided to use International Aero Engines V2500 conventional turbofans instead. IAE is a consortium of Pratt & Whitney, Rolls-Royce, Japan Aero Engines, Fiat, and MTU.

Delta Air Lines provided the MD-90 launch order in November 1989. The first of two prototypes, T-1, made its first flight in February 1993. FAA certification came in November 1994. On 1 April 1995, the MD-90 entered revenue service with Delta.

So far, Douglas has only committed to the MD-90-30, which seats 150-158 in a two-class configuration. It competes with the Boeing 737-300/-700 and the Airbus A320. But Douglas has proposed several variants, including the shrunken MD-90-10 and the stretched MD-90-50. Douglas has also proposed the MD-95, a largely unrelated 105-seat jetliner.

As with the MD-80, some MD-90s will be assembled by China's SAIC. Current plans call for SAIC to assemble at least 20 MD-90s, with first deliveries in 1997.

Specifications (MD-90-30):

Powerplant: two International Aero Engines V2525-D5 turbofans, each rated at 111.21 kN (25,000 lbst)
Dimensions: length: 46.51 m (152 ft 7 in); height: 9.33 m (30 ft 7 in); wing span: 32.87 m (107 ft 10 in)
Weights: empty operating: 40,007 kg (88,200 lb); MTOW: 70,760 kg (156,000 lb)
Performance: cruise speed: 809 km/h (437 kts); range: 4,200 km (2,266 nm)
Passengers: 153 (two class)

McDonnell Douglas MD-95

Douglas' newest narrowbody was launched in 1995

The MD-95 is a narrowbody twinjet designed for short/medium-range routes. Essentially a reincarnation of the DC-9-30, the MD-95 has new avionics and Rolls-Royce/BMW BR715 turbofans. It seats 106 passengers in two classes, or 129 in one class. McDonnell Douglas first announced the MD-95 in 1991, as a joint venture with China's CATIC. Northwest Airlines agreed to evaluate the type, then powered by Pratt & Whitney engines. This Chinese-US manufacturing plan fell through, but Douglas continued to market the aircraft.

In 1994 Douglas began anew, actively offering the

MD-95 to airlines. The company selected numerous contractors for the plane. Dalfort Aviation of Dallas, Texas was selected to assemble the MD-95, but manufacture was later transferred to Douglas's Long Beach, California plant. In late 1994 Douglas began building a static MD-95 prototype, using an ex-Eastern Airlines DC-9-30.

After many false starts, the MD-95 was launched in October 1995. US discount carrier Valujet ordered 50 firm and 50 option aircraft. Valujet flies only used DC-9s, and will buy more until the MD-95 enters service. The MD-95 will make its first flight in May 1998. It will enter service with Valujet in June 1999. Beyond that, Douglas will also offer a stretched MD-95, seating about 122 passengers. There will also be an extended-range variant.

Specifications (MD-95):

Powerplant: two BMW/Rolls-Royce BR 715 turbofans, each rated at 82.29 kN (18,500 lbst)

Dimensions: length: 36.36 m (119 ft 3.5 in); height: 8.60 m (28 ft 2.5 in); wing span: 28.44 m (93 ft 3.5 in)

Weights: empty operating: 30,073 kg (66,300 lb); MTOW: 51,710 kg (114,000 lb)

Performance: cruise speed: 811 km/h (438 kts); range: 2,778 km (1,500 nm)

Passengers: 100 (two class)

Tupolev Tu-134

Unlucky Russian passengers fly the Tu-134B which is extremely cramped with 96 people inside

The Tu-134 is a narrowbody twinjet airliner designed for short/medium-range routes. It was designed with features taken from previous Tupolev Design Bureau jetliners (and bombers), such as the Tu-104 and Tu-124. The Soviet equivalent of the DC-9, BAC 1-11, and Fokker 28, the Tupolev design has engines mounted on the rear fuselage with a T-tail.

Production of the 72-seat Tu-134 began in 1964, and the type entered service in September 1967. The Tu-134 was followed by the 80-seat Tu-134A variant,

which was stretched by 2.1 m (6 ft 11 in). The Tu-134A, also featuring D-30 Series II engines, first appeared in 1970.

Many Tu-134s and -134As were later modified to Tu-134B, B-1, and B-3 configuration, with internal changes. The Tu-134B-3 allowed airlines to cram up to 96 passengers inside.

Tu-134 manufacture ended in the early 1980s. Over 700 were built, including about 100 for export.

Today, most of the ex-Soviet and Eastern Bloc airlines still fly Tu-134s. As of 1995 over 440 were still in service. Big users include Komi Avia (37), Aeroflot Russian (15), Belavia (19), and Air Ukraine (26). The Tu-134 will remain in service until these airlines can afford to replace them with something more fuel-efficient.

Specifications (Tu-134A):

Powerplant: two Soloviev D-30 Series II turbofans, each rated at 65.3 kN (14,990 lbst)

Dimensions: length: 37.05 m (121 ft 6.5 in); height: 9.14 m (30 ft 0 in); wing span: 29.0 m (95 ft 1.75 in)

Weights: empty operating: 29,050 kg (64,045 lb); MTOW: 47,000 kg (103,600 lb)

Performance: cruise speed: 750 km/h (405 kts); range: 1,890 km (1,020 nm)

Passengers: 76 (one class)

Tupolev Tu-154

Almost every Communist airline flew Tu-154s

The Tu-154 was the Soviet Union's answer to the Boeing 727 - a narrowbody medium/long-range jetliner with three rear-mounted engines and a T-tail. Like the 727, the Tu-154 carries about 150 passengers, or a maximum of 180. The Tu-154 can be distinguished by large wing fairings, which house the landing gear. The Tupolev Design Bureau began the Tu-154 programme in 1966 as a replacement for the Tu-104 and Ilyushin Il-18. The first of six prototypes flew in October 1968. The Tu-154 entered service in February 1972 and received the unfortunate NATO

code name 'Careless'.

The first models were the Tu-154, Tu-154A, and Tu-154B. These were powered by increasingly powerful versions of the Kuznetsov NK-8 turbofan. The final production version was the Tu-154M. This features more efficient D-30 turbofans and airframe modifications. A prototype, converted from a Tu-154B-2, flew in 1982. Tu-154M deliveries began in December 1984.

Over 1,000 Tu-154s were built, and the production line at Kuybyshev is still warm today. As of 1995, over 750 were still in service, in the places you would expect to find them - Russia, China, and the Third World. Big users include Aeroflot Russian (30), Kazakhstan Airlines (24), Air Ukraine (31), Vnukovo Airlines (30), and Uzbekistan Airways (25).

Specifications (Tu-154M):

Powerplant: three Aviadvigatel D-30KU-154-II turbofans, each rated at 104 kN (23,380 lbst)

Dimensions: length: 47.90 m (157 ft 1.75 in); height: 11.40 m (37 ft 4.75 in); wing span: 37.55 m (123 ft 2.5 in)

Weights: empty operating: 55,300 kg (121,915 lb); MTOW: 100,000 kg (220,460 lb)

Performance: cruise speed: 950 km/h (513 kts); range: 3,900 km (2,105 nm)

Passengers: 180 (one class)

Tupolev Tu-204

The Tu-204 is now available with Rolls Royce engines

The Tu-204 is a narrowbody twinjet commercial transport designed by the Tupolev Design Bureau. Intended to replace the Tu-154, the Tu-204 seats 190-214 passengers and is about the same size as the Boeing 757 and Airbus A321. The Tu-204 programme began in the early 1980s. The first of six prototypes flew in January 1989. Tu-204 cargo operations began in early 1993, and Russian passenger certification was awarded in early 1995.

Small numbers of Tu-204s have been built at the Ulyanovsk production line. Vnukovo Airlines and Aeroflot Russian (ARIA) have a few each, and the Russian government uses two for state duties. Full

production, as with other CIS planes, is delayed pending the resolution of financial problems.

Like the Il-96 widebody, the Tu-204 uses PS-90 engines. Also like the Il-96, the Tu-204 is available with Western engines and avionics. The Tu-204M uses Rolls-Royce RB.211-535 engines. The avionics, including a two-crew flight deck, are from Rockwell-Collins, Honeywell, and other Western manufacturers. The 204M, also known as the Tu-204-222, began flight tests in 1992. It is marketed by the British-Russian Aviation Company (BRAVIA), but there have been no orders yet.

Tupolev is also planning a 160-seat shortened version, the Tu-234, or Tu-204-300. This was rolled out in August 1995, but has not entered production.

Specifications (Tu-204):

Powerplant: two Aviadvigatel PS-90A turbofans or two Rolls-Royce RB211 535 E4 turbofans.

Data below is for aircraft with PS-90As each rated at 158.3 kN (35,580 lbst)

Dimensions: length: 46.0 m (150 ft 11 in); height: 13.90 m (45 ft 7.25 in); wing span: 42.0 m (137 ft 9.5 in)

Weights: empty operating: 58,300 kg (128,530 lb); MTOW: 94,600 kg (208,550 lb)

Performance: cruise speed: 830 km/h (448 kts); range: 2,900 km (1,565 nm)

Passengers: 196 (two class)

Airtech CN-235

Primarily a military aircraft, the CN-235 is also operated as a commercial airliner

The CN-235 is a pressurized twin-turboprop commuter/utility aircraft. Known primarily as a military aircraft, the CN-235 merits inclusion in this book because it has found its way into several airlines. The CN-235 is built by Aircraft Technology Industries, or Airtech, a joint venture comprising Spain's CASA and Indonesia's IPTN. There are final assembly lines in both countries. CASA builds the forward and centre fuselage, engine nacelles, wing centre section and inboard flaps. IPTN builds the outer wings, ailerons, rear fuselage, and tail.

CN-235 design began in 1980. Each country built a prototype, and both were rolled out in September 1983. Certification and first deliveries came in December 1986. The CN-235 Series 200 is the current civil production version. CASA is also considering a stretched variant with increased cargo capacity. Commercial versions of the CN-235 are only found in service with a handful of airlines. Not surprisingly, the only important users are in Spain and Indonesia. Spain's Binter Canarias and Binter Mediterraneo each have five. Indonesia's Merpati has 14, while Pelita and Dirgantara Air Services have each ordered ten. Airtech has built over 170 CN-235s, and production is continuing. A third assembly line has been set up at Turkey's Tusas plant to fill an order for 50 transports for the Turkish military.

Specifications (CN-235-100):

Powerplant: two General Electric CT7-9C turboprops, each rated at 1,305 kW (1,750 shp)

Dimensions: length: 21.40 m (70 ft 2.5 in); height: 8.177 m (26 ft 10 in); wing span: 25.81 m (84 ft 8 in)

Weights: empty operating: 9,800 kg (21,605 lb); MTOW: 15,100 kg (33,289 lb)

Performance: cruise speed: 460 km/h (248 kts); range: 834 km (450 nm)

Passengers: 44

Antonov An-24/26/32

An Antonov An-24 refuels in Alaska

The An-24/26/32 is a family of twin-turboprop, high-wing regional airliners designed by the USSR's Antonov Bureau. The first model, the An-24, seats 44-50 passengers. The An-26 is a development of the An-24 with a rear-loading ramp. The An-32 is an An-26 with more powerful engines.

An-24 development began in the late 1950s. The An-24 first flew in April 1960. It entered service in September 1963 and was produced in numerous configurations. The An-24RV and RT had an

auxiliary turbojet in the starboard engine nacelle for better take-off performance. The An-24P was designed to paradrop firefighters. The An-30 is an An-24 variant designed for aerial survey missions. The An-26, designed primarily for military roles, appeared in 1970. The An-32, used for hot-and-high operations, first appeared in 1977. It is distinguished by its engines, mounted above the wings.

Production of the An-24/26/32 series ended in the Soviet Union in the late 1970s after over 1,100 aircraft had been built. The An-24 lives on in China, as the Xian Y-7. As of 1995, hundreds of the Antonov propliners were still in service, mostly in the former Soviet Union. Sibavia has over 40. Magadan has 20. Far Eastern Aviation has 25. Romania's Tarom has 19. Cubana has 12.

Specifications (Antonov An-32):

Powerplant: two ZMKB Progress AI-20D Series 5 turboprops, each rated at 3,760 kW (5,042 ehp) **Dimensions:** length: 23.78 m (78 ft 0.25 in); height: 8.75 m (28 ft 8.5 in); wing span: 29.20 m (95 ft 9.5 in)

Weights: empty operating: 17,308 kg (38,158 lb); MTOW: 27,000 kg (59,525 lb)

Performance: cruise speed: 470 km/h (254 kts); range: 1,200 km (647 nm)

Passengers: 50

Antonov An-28

Although designed in Russia and flown by Aeroflot, the Antonov An-28 is manufactured in Poland

The An-28 is a development of the piston-powered An-14, a utility transport which first flew in 1958. Originally called the An-14M, the An-28 twin-turboprop first appeared in September 1969. Flight testing was completed in 1972.

Designed by then Antonov Design Bureau for service on Aeroflot's shortest routes, the An-28 is a rugged, unpressurized aircraft. It is distinguished by its very wide, high, braced wing. The non-retractable landing gear is mounted on sponsons.

The An-28 received Soviet certification in 1978. Under an agreement signed the same year, all An-28 production was moved to PZL Mielec in Poland. PZL deliveries began in 1984.

The An-28 is used for a wide variety of applications. It can carry 17 passengers, but it can also be used for freight, search and rescue, agricultural tasks, and geological survey. The An-28B1T variant is used to airdrop firefighters. An-28 production is continuing and over 180 have been built. The latest version is the An-28RM, a patrol version built for the Polish Navy. For the future, PZL has proposed the An-28PT (or M-27), a version with Pratt & Whitney Canada PT6A turboprops. It could also use Western avionics.

A stretched and modified version of the An-28 is the An-38.

Specifications (An-28):

Powerplant: two PZL Rzeszow PZL-10S (TWD-10B) turboprops, each rated at 716 kW (960 shp)
Dimensions: length: 13.10 m (42 ft 11.75 in); height: 4.90 m (16 ft 1 in); wing span: 22.063 m (72 ft 4.5 in)
Weights: empty operating: 3,900 kg (8,598 lb); MTOW: 6,500 kg (14,330 lb)
Performance: cruise speed: 335 km/h (181 kts); range: 560 km (302 nm)
Passengers: 17

Antonov An-38

A stretched An-28, the An-38 first flew in 1994

The An-38 is a twin-turboprop 26-seat regional airliner designed by Antonov of the Ukraine. Designed for short flights, the An-38 is a stretched version of the An-28. It is unpressurized, with a rear cargo ramp and non-retractable landing gear. Antonov made its An-38 design public in 1991.

The An-38 is notable as the only former Soviet turboprop now available with Western engines and systems. In September 1993 Antonov signed an agreement with AlliedSignal Aerospace to use the latter company's avionics and TPE331 engines on the

An-38. The An-38 will also be available in a cheaper variant, with Russian RKBM Rybinsk TVD-20 turboprops.

The An-38 made its first flight in June 1994, by which time five aircraft were on the production line.

The An-38 is designed to replace An-24s, An-26s, and Let 410s in Russian/Ukrainian airlines. Antonov claims to have 'preliminary' orders for 136 An-38s, including three firm orders from Russia's Vostok Airlines. The An-38 could receive Western certification in early 1996. Deliveries could begin in October 1996.

An-38 production will take place at Russia's NAPO Novossibirsk plant. The aircraft will be marketed by a Russo-Ukrainian joint venture, established by Antonov and NAPO.

Specifications (An-38):

Powerplant: two RKBM/Rybinsk TVD-1500 turboprops, each rated at 1,104 kW (1,480 shp)
Dimensions: 22.06 m (72 ft 4.5 in); height: 4.30 m (14 ft 1.25 in); wing span: 22.06 m (72 ft 4.5 in)
Weights: empty operating: 5,087 kg (11,215 lb); MTOW: 8,800 kg (19,400 lb)
Performance: cruise speed: 350 km/h (188 kts); range: 600 km (323 nm)
Passengers: 26

Avions de Transport Regional ATR 42/72

The ATR42 was the first of the regional aircraft series

The ATR series comprises two twin-turboprop Regional Transport Aircraft (ATR in French and Italian) developed and built by France's Aerospatiale and Italy's Alenia. The ATR 42 seats 42-50, while the stretched ATR 72 seats 70-80.

The aircraft are high-wing pressurized designs with digital avionics. ATRs can be distinguished from similar aircraft (Fokker 50, DHC Dash 8) by their landing gear, which retract inside the fuselage, not the engine nacelles.

In the late 1970s both Aerospatiale and Aeritalia (now Alenia) announced new regional airliners. The two companies decided to pool their efforts in 1980, and in October 1981 they launched the ATR 42 with an order from Air Littoral. The ATR 42 made its first flight in August 1984 and entered service in December 1985. The latest ATR 42 is the faster 42-500, which features the PW127 engines used on the ATR 72.

The ATR 72 programme began in January 1986. The type flew in October 1988, entering service one year later with Finland's Karair. It is also available with uprated engines for hot-and-high operations as the ATR 72-210.

ATR production is continuing, and as of late 1995 ATR had delivered over 285 ATR 42s and 150 ATR 72s. There are over 65 ATR customers, including Simmons (26 42s and 31 72s), Continental Express (42 42s), Eurowings (17 42s and 11 72s), and Transasia (16 72s).

Specifications (ATR 42-300):

Powerplant: two Pratt & Whitney Canada PW120 turboprops, each rated at 1,342 kW (1,800 shp)
Dimensions: length 22.67 m (74 ft 4.5 in); height: 7.586 m (24 ft 10.75 in); wing span: 24.57 m (80 ft 7.5 in)
Weights: empty operating: 10,285 kg (22,674 lb); MTOW: 16,700 kg (36,817 lb)
Performance: cruise speed: 498 km/h (269 kts); range: 1,944 km (1,050 nm)
Passengers: 42

British Aerospace 748

Like the F27, the 748 used Rolls Royce Dart engines

The 748 was designed by Britain's Avro company, later absorbed by Hawker Siddeley and then British Aerospace. A low-wing twin-turboprop pressurized design, the 748 was designed in the late 1950s to compete with the Fokker 27. The first of two prototypes flew in June 1960. Certification came in December 1961 and the type entered service in 1962.

The Series 1 was quickly followed by the Series 2, which flew in November 1961. The Series 2B had a larger wing and other improvements. The Series 2C, first flown in December 1971, featured a large freight door on the side of the fuselage. Final version was the Super 748, a Series 2B with a new flight deck.

Numerous regional airlines purchased the 748. The 748 was also used by military customers. The Indian Air Force bought 72 for transport duties, and the Royal Air Force purchased 31, designated Andover C.Mk1.

A total of 380 748s were built, with production ending in 1987. Of these, 89 were built in India by Hindustan Aeronautics Ltd. Over 100 748s are still in service. And, the 748 lived to see the 1990s - the design was upgraded and reborn as the BAe ATP

Specifications (Super 748):

Powerplant: two Rolls-Royce Dart Mk 552 turboprops, each rated at 1,700 kW (2,280 ehp) **Dimensions:** length: 20.42 m (67 ft); height: 7.57 m (24 ft 10 in); wing span: 31.24 m (102 ft 6 in)

Weights: empty operating: 12,274 kg (27,059 lb); MTOW: 21,092 kg (46,500 lb)

Performance: cruise speed: 454 km/h (245 kts); range: 3,055 km (1,650 nm)

Passengers: 40-58

British Aerospace ATP/Jetstream 61

The ATP was a failed attempt to succeed the 748

A good example of going to the well once too often, the Advanced Turbo Prop (ATP) is BAe's follow-on aircraft to its successful 748. Like the 748, the ATP is a pressurized turboprop regional aircraft. It seats 60-72 passengers and features new engines and avionics.

ATP development began in March 1984 and a prototype flew in August 1986. European JAA certification was awarded in March 1988, and the ATP entered service in April.

From the start, the type was plagued by technical problems. The PW126A engines suffered premature degradation and had to be replaced with PW127s. The propeller hubs leaked oil, and the flight control system was faulty.

In October 1992 BAe transferred the ATP to its Jetstream division, and the aircraft was redesignated Jetstream 61. The J61 name came with several upgrades, including a modernized cabin and improved PW127D engines. No orders were received.

The ATP programme ended in early 1995, when British Aerospace agreed to merge its regional aircraft products with the Aerospatiale/Alenia ATR team. As part of the deal, BAe agreed to cease production of the ATP. Just over 60 were built. Big customers included British Airways and Air Wisconsin. The company still hopes to sell a few remaining ATPs, with the J61 upgrades.

Specifications (ATP):

Powerplant: two Pratt & Whitney Canada PW127D turboprops, each rated at 1,781 kW (2,388 shp) maximum continuous power
Dimensions: length: 26 m (85 ft 4 in); height: 7.59 m (24 ft 11 in); wing span: 30.63 m (100 ft 6 in)
Weights: empty operating: 14,242 kg (31,400 lb); MTOW: 23,678 kg (52,200 lb)
Performance: cruise speed: 437 km/h (236 kts); range: 1,741 km (939 nm)
Passengers: 60-72

British Aerospace Jetstream 31/41

The J41 is a stretched version of the J31

The Jetstream 31 is a 19-seat twin-turboprop pressurized regional airliner built by British Aerospace. The Jetstream 41 is its stretched sibling, which accommodates 27-29 passengers.

The J31 design began as the Handley Page Jetstream 1. This first flew in August 1967, with first deliveries in June 1969. Handley Page built about 26 for military applications.

In 1978 successor company BAe decided to develop the design, and a modified Jetstream 1 flew in March 1980. Deliveries began in December 1982. In 1988

production switched to the Super 31, or J32. The J32 features an upgraded interior, increased take-off weight, and uprated engines.

BAe has built over 380 J31/32s, and over 360 of these are still in service. Major users include AMR's Nashville Eagle (60), Chaparral (20), Trans States (22), Jetstream International (31), Express Airlines (25), and Air Midwest (15). Over 300 J31s are being leased by BAe's Jetstream leasing company, JSX.

BAe first revealed the J41 in 1988. Launched in May 1989, the J41 first flew in September 1991, with certification and first deliveries in November 1992. Major J41 users include Atlantic Coast, Manx Airlines, and Trans States.

Production of the J41 is continuing, and BAe is still offering the J31/32 as well.

Specifications (Jetstream 41):

Powerplant: two AlliedSignal/Garrett TPE331-14GR/HR turboprops, each rated at 1,230 kW (1,650 shp)
Dimensions: length: 19.25 m (63 ft 2 in); height: 5.74 m (18 ft 10 in); wing span: 18.29 m (60 ft 0 in)
Weights: empty operating: 6,416 kg (14,144 lb); MTOW: 10,886 kg (24,000 lb)
Performance: cruise speed: 547 km/h (295 kts); range: 1,433 km (774 nm)
Passengers: 29

CASA C-212

Designed by CASA, the C212 is made in Indonesia

The C-212 Aviocar is an unpressurized twin-turboprop utility and passenger transport designed by Spain's CASA. It is also built under licence by Indonesia's IPTN. Uses include airline passenger and freight operations, military transport, and maritime surveillance.

CASA designed the C-212 in the late 1960s as a replacement for Spain's aging DC-3 and JU-52 transports. The first prototype flew in March 1971. Production deliveries began in May 1974. In 1984 CASA introduced the current production model, the

C-212-300. It has uprated engines, a larger interior, and higher take-off weight. CASA is considering a new version, the C-212-400, with digital avionics.

As an airliner, the C-212 seats 26 passengers, or 24 and a lavatory. The freighter version can carry 2,700 kg (5,952 lbs) of cargo. Most C-212 civil operators are Indonesian carriers. Merpati has 15, and Pelita has 12. If you're going to study Orangutans in Sumatra, odds are you will go in a C-212.

Most C-212s are powered by TPE331 turboprops, but CASA also offers the C-212P, powered by Pratt & Whitney Canada PT6A-65Bs. No orders have been received for this model.

CASA has built over 340 C-212s, while IPTN has built over 100 NC-212s. Production is continuing at both lines.

Specifications (C-212 Series 300):

Powerplant: two AlliedSignal/Garrett TPE331-10R-513C turboprops, each rated at 671 kW (900 shp)

Dimensions: length: 16.15 m (52 ft 11.75 in); height: 6.60 m (21 ft 7.75 in); wing span: 20.28 m (66 ft 6.5 in)

Weights: empty operating: 4,400 kg (9,700 lb); MTOW: 7,700 kg (16,975 lb)

Performance: cruise speed: 300 km/h (162 kts); range: 440 km (237 nm)

Passengers: 26

De Havilland Canada DHC-6 Twin Otter

With floats, skis or wheels, the DHC-6 Twin Otter can land almost anywhere

If you find yourself on an expedition to a Mayan temple in Guatemala, flying low over the jungle to a dubious-looking airstrip, there is a an excellent chance that you are in this plane. You are in good hands and will not wind up as some python's lunch. DHC built the Twin Otter as a multi-purpose transport for operations in all sorts of rough conditions. Short

landing fields are not a problem, and the type has also been built with floatplanes and wheel/ski landing gear.

A twin-turboprop, high-wing, unpressurized design, the Twin Otter can carry 20 passengers, or freight. Militaries in the US, Canada, and other countries use the type for dozens of roles. The first version was the Series 100, distinguished by a short nose. The Series 200 and 300 (the final production version) were externally identical, but the 300 had uprated engines.

The Twin Otter first flew in May 1965 at DHC's Downsview, Ontario plant. Certification and first deliveries came one year later. Production totalled 844 when the programme ended in 1988. Due to the aircraft's rugged construction, many of these will still be in service in the next century.

Specifications (DHC-6 Series 300):

Powerplant: Two Pratt & Whitney Canada PT6A-27 turboprops, each rated at 462 kW (620 shp) for take-off

Dimensions: length: 15.8 m (51 ft 9 in); height: 5.9 m (19 ft 6 in); wing span: 19.8 m (65 ft)

Weights: empty operating: 3,363 kg (7,415 lb); MTOW: 5,670 kg (12,500 lb)

Performance: cruise speed: 338 km/h (182 kts); range: 1,297 km (700 nm)

Passengers: 19-20

De Havilland Canada DHC ('Dash') 7

The four-engined DHC-7 is optimised for short field take-offs and landings

For years, DHC was the Hudson Bay Outfitters of aircraft. The DHC-5 was a tough military transport. The Dash-6 was ideal for service in an earthquake zone. Finally, before turning to conventional passenger transports with the Dash-8, DHC built the Dash 7. A four-engine transport, the Dash 7 has enough installed engine power to operate from Short Take-Off and Landing (STOL) runways only 685 m (2,160 ft) in length.

The Dash 7 can seat up to 54 passengers, or cargo, or a mix. The Dash 7 programme began in 1972, and the type made its first flight in March 1975. Canadian certification was awarded in May 1977.

First production version was the Series 100, followed by the heavier Series 150. Cargo versions of both were the Series 101 and 151, respectively. DHC proposed further versions, including the Series 300, a 70-seat stretch. These were cancelled, partly due to the advent of twin-engine transports with the same capacity, such as the ATR 72.

DHC built 111 Dash 7s, with production ending in the late 1980s. Most of these are still in service. To see one, go to an airport in the mountains or in the city. The UK's Brymon and the US's Continental Express and Trans World Express still operate the type.

Specifications (DHC-7 Series 100):

Powerplant: Four Pratt & Whitney Canada PT6A-50 turboprops, each flat-rated at 835 kW (1,120 shp) for take-off
Dimensions: length: 24.54 m (80 ft 6 in); height: 7.98 m (26 ft 2 in); wing span: 28.35 m (93 ft)
Weights: empty operating: 12,560 kg (27,690 lb); MTOW: 19,958 kg (44,000 lb)
Performance: cruise speed: 399 km/h (215 kts); range: 2,168 km (1,170 nm)
Passengers: 50-54

De Havilland Canada DHC-8

Norway's Wideroe uses the 36-seat DHC-8-100

The Dash 8 is the latest in a long line of DHC's Canadian-built turboprop transports. Unlike the Dash 5, the Dash 8 is built for regional airline operations from normal airports. A high-wing, pressurized, twin-engine design, the Dash 8 comes in two basic versions, the 30-36 seat Dash 8-100 and the stretched 50-56 seat Dash 8-300.

Originally known as the Dash X, the Dash 8 programme began in 1978. The Dash 8-100 was rolled out in April 1983. It entered service in

December 1984. DHC also offers the Dash 8-200, a faster -100 with greater commonality with the -300.

The -300 was first announced in mid 1985. It entered service in March 1989. It uses uprated PW123B engines.

Major -100 users include USAir Express, Norway's Wideroe, Air Ontario, Horizon Air, and Northwest Airlines. Canada's military uses the -100, designated CC-142, for passenger and cargo transport. Big -300 users include Time Air and Air Wisconsin. DHC has built over 300 -100s and over 100 -300s. Production is continuing.

DHC has also proposed the Dash 8-400, a further stretch version with new engines. It will carry 70 passengers at 648 km/h (350 kt) speeds. The -400 could be delivered as early as 1998.

Specifications (DHC-8-100):

Powerplant: two Pratt & Whitney Canada PW120A turboprops, each rated at 1,491 kW (2,000 shp) with automatic power reserve
Dimensions: length: 22.25 m (73 ft); height: 7.49 m (24 ft 7 in); wing span: 25.91 m (85 ft)
Weights: empty operating: 10,251 kg (22,600 lb); MTOW: 15,650 kg (34,500 lb)
Performance: cruise speed: 439 km/h (237 kts); range: 1,519 km (820 nm)
Passengers: 30-36

Dornier 228

The Dornier Do 228 is now used by several small airlines

The Do 228 is a 15-19 seat unpressurized twin-turboprop transport used for airline and utility operations. Capable of Short Take-Off and Landing (STOL) operations, the 228 is built by Dornier, now part of Daimler Benz Aerospace.

Design work on the 228 began in the mid 1970s with Dornier's work on its new-technology TNT wing. A prototype 228 flew in March 1981. The type entered service in August 1982.

The baseline 228 is the 15-seat -100. The

lengthened fuselage 228-200 carries 19. The latest production model of the 228 is the -212, which features uprated engines. The -212 was certified in July 1990. The 228 is also available in maritime patrol, sensor platform, cargo, ambulance, and other variants.

Dornier has delivered over 220 228s. About 75 of these are used by governments and corporations. Numerous airlines use small numbers of 228s. Only two - Taiwan's Formosa and the US's Precision - have ordered more than 10.

The Do 228 is also built under licence in India by Hindustan Aeronautics Ltd. HAL has built about 20 aircraft so far. Six have gone to Vayudoot airlines, and the rest have gone to the Indian government and military services.

Specifications (Do 228-212):

Powerplant: two AlliedSignal/Garrett TPE331-5-252D turboprops, each rated at 578.7 kW (776 shp)
Dimensions: length: 16.56 m (54 ft 4 in); height: 4.86 m (16 ft); wing span: 16.97 m (55 ft 8 in)
Weights: empty operating: 3,258 kg (7,183 lb); MTOW: 6,400 kg (14,110 lb)
Performance: cruise speed: 408 km/h (220 kts); range: 1,167 km (630 nm)
Passengers: 19

Dornier 328

After a difficult development programme, the Dornier Do 328 has now entered service and is attracting orders

Dornier's 328 is a twin-turboprop 30-33 seat pressurized regional airliner. Capable of 620 km/h (335 kt) speeds, the 328 is the fastest plane in its class.

Dornier, now part of Daimler Benz Aerospace, began research on a new 30-seat plane in 1984. The 328 project was launched and the design frozen in mid 1989. This was followed by some hard times. The launch customer, Contact Air, cancelled its order in March 1991. A 328 prototype flew in late 1991, but flight tests were stopped to re-engine the plane

with more powerful PW119Bs. Another major customer, Midway, was liquidated in March 1992. Later that year a 328 prototype suffered a near-catastrophic propeller failure.

Still, Dornier pushed on. JAA certification came in October 1993, followed by first delivery to Air Engiadina. The future now looks bright, with a large order from Horizon Air. Horizon will use the type on its US and Canadian Pacific Northwest routes. Corning has ordered two corporate variant 328s.

The 328 is built by an international team of subcontractors. The fuselage is built by South Korea's Daewoo and Italy's Aermacchi. Israel Aircraft Industries builds the wings. Britain's Westland builds the engine nacelles and fuselage doors. The engines, of course, come from Pratt & Whitney Canada.

Specifications (Do 328):

Powerplant: two Pratt & Whitney Canada PW119B turboprops, each rated at 1,380 kW (1,850 shp) for normal take-off
Dimensions: length: 21.22 m (69 ft 8 in); height: 7.24 m (23 ft 9 in); wing span: 20.98 m (68 ft 10 in)
Weights: empty operating: 8,175 kg (18,022 lb); MTOW: 13,640 kg (30,071 lb)
Performance: cruise speed: 620 km/h (335 kts); range: 1,556 km (840 nm)
Passengers: 30-33

Douglas DC-3

The ancestor of all modern airliners, the DC-3 still soldiers on today

This plane helped start it all. The DC-3 helped make airline operations profitable, and ushered in the modern age of air transport. Its important technical innovations, such as retractable landing gear, are eclipsed by its beautiful and instantly recognizable art deco design features. It seats 28-36 passengers, four abreast.

Sixty years of DC-3 history in a paragraph: development began in 1932, with a first flight in 1935. Certification and service entry (with American Airlines) came in 1936. Total production: 10,926, excluding Soviet copies. Of these, over 10,000 were military transports, designated C-47 Skytrain in the US and Dakota in the RAF. If you were a parachutist

in World War II, you probably dropped out of one. As militaries shed these after the war, they became workhorses in nearly every airline.

Actually, the DC-3 had equally important contemporaries, including the smaller DC-2 (from which the DC-3 is derived) and Boeing's 247. But the DC-3 has achieved near immortality. Unlike the others, several hundred DC-3s still soldier on in military and commercial service.

Astute readers may notice that this book covers only turbine-powered aircraft; the DC-3 makes the cut because there are several turboprop retrofit programmes available. The South African Air Force has modified its 27 aircraft with Pratt & Whitney Canada PT6As. Commercial DC-3 users include Canada's Skycraft and Air Manitoba.

Specifications (DC-3):

Powerplant: Two Pratt & Whitney R-1830-92 Twin Wasp radial piston engines, each rated at 895 kW (1,200 hp) for take-off
Dimensions: length: 19.66 m (64 ft 6 in); height: 5.16 m (17 ft); wing span: 28.96 m (95 ft)
Weights: empty operating: 8,030 kg (17,720 lb); MTOW: 11,430 kg (25,200 lb)
Performance: cruise speed: 266 km/h (143 kts); range: 2,430 km (1,312 nm)
Passengers: 28-36

Embraer EMB-110

The unpressurised EMB-110 was Embraer's first project

The Embraer EMB-110 Bandeirante is a twin-turboprop transport used for a variety of civil and military applications. An unpressurized, low-wing design with retractable landing gear, the EMB-110 was the first transport aircraft built in Brazil. It can be configured as a cargo plane, trainer, or as an 18-21 seat regional airliner.

The EMB-110 was developed by the government's Institute for Research and Development (IRD). The first of three prototypes flew in October 1968.

After that, the story of the Bandeirante is also the story of Embraer. The company was created in 1969 to build the EMB-110, primarily for the Brazilian Air

Force. The service received its first EMB-110 in 1973 and ultimately bought about 140 aircraft. The type was also built for maritime patrol as the EMB-111.

The final production Bandeirantes were the EMB-110P1A and P2A. First delivered in December 1983, they featured a Collins electronic flight instrumentation system (EFIS) and a modified tailplane.

The EMB-110 line closed in the late 1980s after production of over 475 aircraft. Embraer planned to move on to its CBA-123 19-seat turboprop as an EMB-110 replacement, but this project collapsed in the early 1990s.

As of 1995 over 60 EMB-110s were in use as airliners. Brazil's TAM has nine. Australia's Flight West has eight. Britain's BAC Express has six.

Specifications (Embraer EMB-110P2):

Powerplant: two Pratt & Whitney Canada PT6A-34 turboprops, each rated at 559 kW (750 shp)

Dimensions: length: 15.1 m (49 ft 6.5 in); height: 4.92 m (16 ft 1.75 in); wing span: 15.3 m (50 ft 3.5 in)

Weights: empty operating: 3,516 kg (7,751 lb); MTOW: 5,670 kg (12,500 lb)

Performance: cruise speed: 335 km/h (181 kts); range: 2,001 km (1,080 nm)

Passengers: 21

Embraer EMB-120

The Brasilia is mainly used by US operators

The EMB-120 Brasilia is a 30-seat twin-turboprop regional airliner built by Brazil's Embraer. A low-wing pressurized design with retractable landing gear and digital avionics, the Brasilia competes with the DHC Dash 8-100, BAe's Jetstream 41, and the Saab 340.

The EMB-120 programme began in the early 1980s. A prototype flew in July 1983. Brazil's CTA certified the EMB-120 in May 1985, and the type entered service in August 1985.

In late 1986 Embraer introduced a hot-and-high

variant, with uprated PW118A engines. Embraer also builds the EMB-120QC, a convertible passenger/cargo aircraft, and the EMB-120ER (Enhanced Range). Finally, in 1994 Embraer introduced the EMB-120ER Advanced, with quieter propellers, an improved interior, and other upgrades.

Embraer has also used the Brasilia fuselage as the basis for two new planes. The EMB-145 regional jet uses a stretched EMB-120 body. The CBA-123 was a 19-seat turboprop with rear-facing engines. Currently shelved, the project used a shortened EMB-120 body.

As of late 1995 Embraer had built almost 300 Brasilias, and production is continuing. Most EMB-120s have gone to North American regional carriers. Big users include Comair, Atlantic Southeast, Continental Express, Skywest, and Westair.

Specifications (Embraer EMB-120ER):

Powerplant: two Pratt & Whitney Canada PW118 turboprops, each rated at 1,342 kW (1,800 shp)

Dimensions: length: 20.07 m (65 ft 10.25 in); height: 6.35 m (20 ft 10 in); wing span: 19.78 m (64 ft 10.75 in)

Weights: empty operating: 7,150 kg (15,763 lb); MTOW: 11,990 kg (26,433 lb)

Performance: cruise speed: 555 km/h (300 kts); range: 1,575 km (850 nm)

Passengers: 30

Fairchild Metro

The Metro's 'flying cigar' profile is very distinctive

The Fairchild Metro is a 19-20 seat pressurized twin-turboprop used for regional airline and utility operations. It competes with Beech's 1900. Created by Ed Swearingen in the 1960s, the Metro is also known as the SA227. The design is distinguished by its narrow 'flying cigar' fuselage. Fairchild Industries purchased the design in 1972.

The original Metro made its first flight in August 1969 and entered service in July 1970. It was followed by the Metro II in 1974 and the Metro III,

introduced in 1980. Oddly, the Metro II had an optional small rocket unit in its tail to assist with take-offs in hot-and-high conditions. Current production model is the Metro 23, a Metro III variant.

As of late 1995 over 950 Metros had been built. The Metro is primarily used by regional airlines such as Aeromexico and Austria's Eurosky. A corporate version is known as the Merlin 23. A cargo variant, called Expediter, is used by UPS and DHL.

The US military uses the Metro, designated C-26, for transport, cargo, and medical evacuation missions. Fairchild and Lockheed Martin are marketing the Multi-Mission Surveillance Aircraft (MMSA), a Metro variant with various sensors for military reconnaissance.

Specifications (Metro 23):

Powerplant: two AlliedSignal Aerospace/Garrett TPE331-12UA-UAR-701G turboprops, each rated at 820 kW (1,100 shp)
Dimensions: length: 18.09 m (59 ft 4 in); height: 5.08 m (16 ft 8 in); wing span: 17.37 m (57 ft)
Weights: empty operating: 4,309 kg (9,500 lb); MTOW: 7,484 kg (16,500 lb)
Performance: cruise speed: 542 km/h (293 kts); range: 2,065 km (1,114 nm)
Passengers: 19-20

Fokker 27

With 786 built, the F27 programme is the most successful European transport programme to date

The F27 Friendship was designed by Fokker in the early 1950s as a twin-turboprop, pressurized high-wing transport. Intended to replace the Douglas DC-3, the programme was launched with Netherlands government support in 1953. The first of two prototypes flew in November 1955. The 44-seat F27 Mk 100 entered service in November 1958.

To help penetrate the US market, Fokker conducted a licence production agreement with Fairchild. Between 1958 and 1970, Fairchild built the F27 and a stretched variant, the FH-227, in Hagerstown, Maryland. Fairchild also built a corporate version, the F27F. Fokker built numerous variants of the F27. The

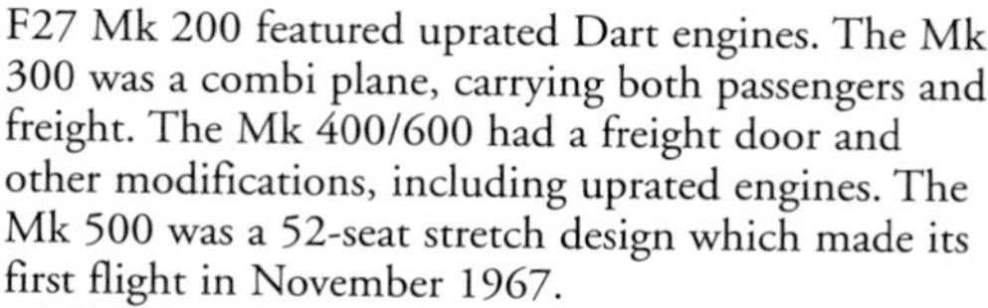

F27 Mk 200 featured uprated Dart engines. The Mk 300 was a combi plane, carrying both passengers and freight. The Mk 400/600 had a freight door and other modifications, including uprated engines. The Mk 500 was a 52-seat stretch design which made its first flight in November 1967.

Military F27s include the F27M Troopship, a military transport, and the F27 Maritime and Maritime Enforcer, two naval patrol variants.

Fokker ended the F27 programme in early 1987. With production totalling 579 Fokker-built and 128 Fairchild-built aircraft, the F27 was the most successful post-war European turboprop transport to date. Like its competitor, the British Aerospace 748, the F27 was reborn with new technology, as the Fokker 50.

Specifications (F27 Mk 200):

Powerplant: two Rolls-Royce Dart Mk 522 turboprops, each rated at 1,700 kW (2,280 ehp)

Dimensions: length: 23.56 m (77 ft 4 in); height: 8.51 m (27 ft 11 in); wing span: 29.0 m (95 ft 2 in)

Weights: empty operating: 11,159 kg (24,600 lb); MTOW: 20,410 kg (45,000 lb)

Performance: cruise speed: 480 km/h (259 kts); range: 2,211 km (1,193 nm)

Passengers: 44

Fokker 50

The Fokker 50 and 60 were developed from the F27

Fokker developed the F50 as a modernized follow-on to its F27. The F50 is about the same size as its forebear, but features new engines, avionics, and other systems. The Rolls-Royce Dart engines have been replaced with Pratt & Whitney Canada PW125Bs, and there is an electronic flight instrumentation system (EFIS).

The F50 programme began in November 1983, and an F50 first flew in December 1985. Launch customer DLT, now Lufthansa Cityline, received its first F50 in August 1987.

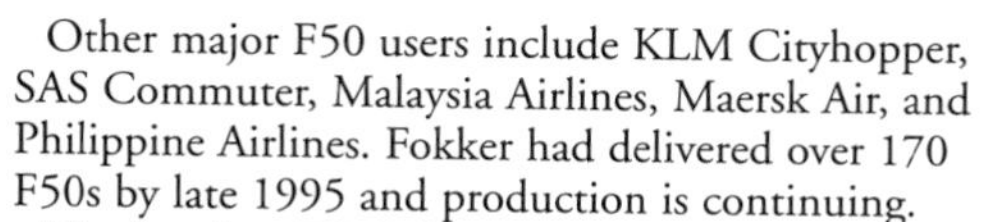

Other major F50 users include KLM Cityhopper, SAS Commuter, Malaysia Airlines, Maersk Air, and Philippine Airlines. Fokker had delivered over 170 F50s by late 1995 and production is continuing.

The baseline F50 is known as the Series 100. Also available is a high performance variant with PW127B turboprops. Fokker is building a maritime patrol version of the F50 known as the Maritime Enforce Mk.2 and offers a variety of other military F50 variants.

Fokker is also building the Fokker 60, a 58 seat stretch version ordered by the Netherlands Air Force. It features a large cargo door on the starboard side of the fuselage. The F60 will be delivered in March 1996 and is also on offer as a civil passenger and freight aircraft.

Specifications (Fokker 50-100):

Powerplant: two Pratt & Whitney Canada PW125B turboprops, each flat-rated at 1,864 kW (2,500 shp)
Dimensions: length: 25.25 m (83 ft 10 in); height: 8.32 m (27 ft 4 in); wing span: 29 m (95 ft 2 in)
Weights: empty operating: 12,520 kg (27,602 lb); MTOW: 19,950 kg (43,980 lb)
Performance: cruise speed: 522 km/h (282 kts); range: 2,253 km (1,216 nm)
Passengers: 50

Harbin Y-12

The robust Y-12 is used by several small airlines

The Y-12 (or Yun-12) is a twin-turboprop utility aircraft built in China by Harbin Aircraft Manufacturing Corporation. If you think the Pilatus Britten-Norman Islander or DHC-6 Twin Otter are too luxurious, you will love the Y-12. A rugged, unpressurized design with braced, high wings and non-retractable landing gear, the Y-12 is as Spartan as they come. Common uses for the Y-12 include passenger transport (it seats 18-19), cargo, and crop spraying.

A Y-12 with Chinese engines made its first flight in 1982, but the current configuration, with Pratt &

Whitney Canada engines, did not fly until August 1984. This model received Chinese certfication in late 1985.

As of late 1995 Harbin had delivered over 80 Y-12s. About 20 of these have gone to Chinese operators, such as China Southwest. Most of the 60 aircraft built for export have gone to Third World carriers, such as Mongolian Airlines, Nepal Airlines, and Lao Aviation.

While the Y-12 is powered by Western engines and other systems, there have been few exports to the West, mostly because the plane has not been FAA certified. But Harbin is promoting a new version, the Y-12-4, which has new wingtips, brakes, and landing gear. This may obtain FAA certification in the next few years.

Specifications (Y-12):

Powerplant: two Pratt & Whitney Canada PT6A-27 turboprops, each rated at 462 kW (620 shp)
Dimensions: length: 14.86 m (48 ft 9 in); height: 5.27 m (17 ft 3.75 in); wing span: 17.23 m (56 ft 6.5 in)
Weights: empty operating: 3,000 kg (6,614 lb); MTOW: 5,300 kg (11,684 lb)
Performance: cruise speed: 240 km/h (129 kts); range: 410 km (221 nm)
Passengers: 17

Ilyushin Il-114

Future Il-114s may include western avionics

The Il-114 is a pressurized twin-turboprop 60-70 seat transport designed by the Ilyushin Design Bureau. A low-wing design, the Il-114 is about the same size and configuration as British Aerospace's ATP.

Seeking to create a replacement for the Antonov An-24, Ilyushin began work on the Il-114 in 1985. The type is powered by Klimov TV7 engines and uses digital avionics.

The first prototype flew in March 1990. One of the five prototypes crashed in July 1993, but Ilyushin pressed on. Eight aircraft have been sent to Moscow for certification and training.

Small numbers of Il-114s have been built at the main production line in Tashkent, Uzbekistan. Production has been held up by financial problems now plaguing most of the ex-Soviet aerospace industry. Still, Ilyushin claims to hold over 80 orders for the Il-114, including ten for Uzbekistan Airways and five for Archangelsk Civil Aviation.

In 1990, Ilyushin held talks with Spain's CASA about using the Il-114 fuselage as the basis for a new airliner, the CASA 3000. This went nowhere, but in the future there could be a different Westernized version of the Il-114 with Pratt & Whitney Canada engines.

Ilyushin has also proposed a shortened 32-seat version of the Il-114, designated Il-112. If it goes ahead, it would replace Let L-610s in Russian service.

Specifications (Il-114):

Powerplant: two Klimov TV7-117 turboprops, each rated at 1,839 kW (2,466 shp)
Dimensions: length: 26.87 m (88 ft 2 in); height: 9.32 m (30 ft 7 in); wing span: 30.00 m (98 ft 5.25 in)
Weights: empty operating: 15,000 kg (33,070 lb); MTOW: 22,700 kg (50,045 lb)
Performance: cruise speed: 500 km/h (270 kts); range: 1,000 km (540 nm)
Passengers: 74

IPTN N-250

The N-250 regional transport is IPTN's first solo design

Indonesia's IPTN is designing the N-250 as a 64-68 passenger twin-turboprop regional airliner. More than just a plane, the N-250 is the culmination of Indonesia's techno-nationalist dreams. Its father, IPTN's Dr. B.J. Habibie, has been steadily building the country's aerospace industry over the last 20 years, and this will be its first indigenous aircraft.

While purportedly a local design, the N-250 owes much to the early CASA/IPTN CN-235. Also, most N-250 subsystems, including the engines, propellers, avionics, landing gear, and other systems, will come

from the US and Europe. IPTN will 'bend metal' and put the plane together.

Habibie first announced the N-250, then a 50-seat design, at the Paris Air Show in June 1989. At the 1993 Paris show, IPTN revealed that the N-250 had grown into its present 64-68 seat size. The pressurized aircraft will use Allison AE2100 engines and state-of-the-art Collins Pro Line 4 avionics.

The first N-250 was rolled out at IPTN's Bandung plant in November 1994. If Habibie stays in power, N-250 production could begin in 1997. Unsurprisingly, most N-250 orders have come from Indonesian carriers. Bouraq has ordered 62, and Merpati Nusantara has ordered 65. IPTN wants to build more N-250s at a second production line in the US, assuming sufficient local demand.

Specifications (N-250-100):

Powerplant: two Allison AE 2100C turboprops, each rated at 2,622 kW (3,517 shp)
Dimensions: length: 28.153 m (92 ft 4.5 in); height: 8.785 m (28 ft 9.75 in); wing span: 28.00 m (91 ft 10.25 in)
Weights: empty operating: 13,665 kg (30,126 lb); MTOW: 22,000 kg (48,502 lb)
Performance: cruise speed: 556 km/h (300 kts); range: 1,482 km (800 nm)
Passengers: 64

Let L-410

Aeroflot bought some 885 L-410s from Czechoslovakia

The L-410 Turbolet is an unpressurized 19-seat twin-turboprop transport designed and built by Let Aeronautical Works in the Czech Republic. The L-410 is a high-wing design, primarily used for passenger operations.

The original prototype for the L-410 was the XL-410. It first flew in April 1969. The L-410A entered service with Slov-Air in late 1971. The L-410A used Pratt & Whitney Canada PT6A-27 engines. Let built 31 L-410As.

In 1973 the L-410A was replaced by the L-410M, which was the first application for the new Czech Motorlet M601 turboprop. Looking back from a time when Eastern European airframers are desperately trying to get Western engines for their planes, the L-

410M can be seen as a great step backward.

The final production model was the L-410UVP-E, which flew in late 1984. This version has uprated engines and interior improvements. Externally, it is distinguishable by wingtip fuel tanks.

L-410 production ended in the early 1990s after nearly 1,050 were built. About 885 went to Aeroflot, and most of these are still in service. Other users include Bolivia's AeroSur and other Third World airlines.

The L-410 could be revived as the L-420. This version features improved M601F engines, new interiors, and other upgrades. General Electric is cooperating on the engine improvements. Let hopes to obtain Western certification for the L-420, but there are no orders yet.

Specifications (L-410UVP-E):

Powerplant: two Motorlet (Walter) M601E turboprops, each rated at 559 kW (750 shp)
Dimensions: length: 14.42 m (47 ft 4 in); height: 5.83 m (19 ft 2 in); wing span: 19.98 m (65 ft 7 in)
Weights: empty operating: 4,160 kg (9,171 lb); MTOW: 6,400 kg (14,110 lb)
Performance: cruise speed: 365 km/h (197 kts); range: 1,357 km (744 nm)
Passengers: 19

Lockheed Martin L-100

Based on the widely-used C-130 military transport, the L-100 has been flying for 30 years

The L-100 is the civilian version of Lockheed's popular C-130 Hercules military transport. A high-wing four-turboprop design, the L-100 can lift cargo or 80-120 passengers. It has internal cargo handling facilities and a four-man flight crew.

An L-100 demonstrator first flew in April 1964 and received FAA certification in February 1965, although the C-130 series began flying in 1954. Company designation for the L-100 is the Model 382.

Some 115 L-100s have been built, mostly for cargo applications. Of these, 29 are L-100-20s (stretched

2.5 metres from the original L-100) and 78 are L-100-30s (4.5 metres longer than the L-100). These figures are dwarfed by the 2,000-plus C-130 military planes built so far.

Major L-100 users include Southern Air Transport in the US and South Africa's Safair Freighters. The Saudi Government has five L-100-30HSs, which are self-contained flying hospitals complete with operating theatres. Aside from these three operators, most of the 25 or so L-100 users have only 1-4 aircraft. For the future, Lockheed Martin is developing the L-100J, with new Allison AE2100D3 engines, a two-man cockpit, and an optional side cargo door. Externally similar to the L-100-30, the L-100J will enter service in the late 1990s, along with the military C-130J.

Specifications (L-100-30):

Powerplant: four Allison Engines 501-D22A turboprops, each rated at 3,490 kW (4,680 ehp).
Dimensions: length: 34.37 m (112 ft 9 in); height: 11.66 m (38 ft 3 in); wing span: 40.4 m (132 ft 7 in)
Weights: empty operating: 35,235 kg (77,680 lb); MTOW: 70,308 kg (155,000 lb)
Performance: cruise speed: 583 km/h (315 kts); range: 2,526 km (1,363 nm)
Payload: 23,158 kg (51,054 lb) or 80-120 passengers

Raytheon 1900

The 1900 is one of the few turboprops built in the USA

The 1900 is a 19-seat pressurized twin-turboprop built by Raytheon's Beech unit. Beech began 1900 development in 1979 as a follow-on to the Beechcraft C99, a 15-seat transport which ended production in 1975. The 1900 first flew in September 1982 and entered service in February 1984. The 1900 airframe is based on the Super King Air 200.

Most 1900s are used by regional carriers, especially Mesa Airlines. Mesa has ordered over 100 of the type, using them in its United Express, Air Midwest, and

Skyway divisions.

It is also used as a business transport, most notably by Mobile Corporation. Several air forces use the type for transport duties, and US Air National Guard has designated them C-12J for electronic surveillance missions.

Beech built 255 1900s (mostly 1900Cs) before switching to the current production model, the longer-ranged 1900D. It entered service in late 1991 and is externally distinguished by winglets. It also has a 'wet' wing (with integral fuel tanks instead of bladders) and an electronic flight instrumentation system (EFIS). Most important, the 1900D centre ceiling is 35.6 cm (14 in) higher than earlier 1900s, so you can stand up inside the cabin without banging your head.

Specifications (1900D):

Powerplant: two Pratt & Whitney Canada PT6A-67D turboprops, each flat-rated at 954 kW (1,279 shp)
Dimensions: length: 17.63 m (57 ft 10 in); height: 4.57 m (15 ft); wing span: 17.67 m (57 ft 12 in)
Weights: empty operating: 4,785 kg (10,550 lb); MTOW: 7,688 kg (16,950 lb)
Performance: cruise speed: 533 km/h (288 kts); range: 2,776 km (1,498 nm)
Passengers: 19

Saab 340

The Saab 340 is popular with both European and North American operators

The 340 is a 30-37 seat regional airliner. Built by Sweden's Saab, the 340 was originally designed with Fairchild Industries as the SF 340A. Fairchild left the joint venture in 1985, and the SF 340 became the Saab 340.

The programme began in 1980, and a prototype first flew in January 1983. The 340 entered service in June 1984. The 340B is the current production model. It first flew in April 1989 and features uprated CT7 engines and improved range and payload.

A twin-turboprop pressurized design, the 340 is

available in cargo and executive variants. Saab proposed a stretched variant, and this became the nucleus of the Saab 2000.

The 340 recently found a new role as a radar-equipped airborne early warning platform. The Swedish Air Force ordered five of these in 1993. This 340AEW is recognizable by the large rectangular Erieye radar carried above the fuselage.

As of late 1995 Saab has built over 350 340s, mostly for European and North American operators. Major users include AMR's Eagle network, Business Express, Northwest Airlink, Crossair, Swedair, and Comair. The 340 competes with the Do.328, EMB-120, Jetstream 41, and DHC-8-100. Production is continuing, and Saab is planning upgrades necessary to keep the 340 competitive into the next century.

Specifications (Saab 340B):

Powerplant: two General Electric CT7-9B turboprops, each rated at 1,305 kW (1,750 shp) for take-off

Dimensions: length: 19.73 m (64 ft 9 in); height: 6.97 m (22 ft 11 in); wing span: 21.44 m (70 ft 4 in)

Weights: empty operating: 8,140 kg (17,945 lb); MTOW: 13,155 kg (29,000 lb)

Performance: cruise speed: 467 km/h (252 kts); range: 1,732 km (935 nm)

Passengers: 35

Saab 2000

Regional airlines and Crossair both fly the Saab 2000

A twin-turboprop regional transport, the Saab 2000 is derived from the Saab 340. The 2000 has a longer fuselage, seating 50-58 passengers, and 33% larger wings than the 340. It also has larger Allison AE2100 engines, giving the 2000 superior range and speed, or 'near-jet performance' as Saab says.

Thin and sleek, the 2000 is also a head-turner. It features six-bladed slow-revving swept propellers. The passenger doors are compatible with standard jetways, eliminating the dreaded passenger shuttle to the plane on the tarmac.

The Saab 2000 will need these attractions. It is competing for sales with the new regional jets, particularly Canadair's RJ. It also competes with established prop designs like the Fokker 50 and the ATR 42.

Saab began the 2000 programme in December 1988. The first of three test aircraft flew in March 1992. The Saab 2000 received European JAA certification in March 1994 and US FAA certification one month later. Switzerland's Crossair received the first of 20 Saab 2000s in August 1994.

In addition to Crossair, the 2000 has been ordered by Deutsche BA and Air Marshall Islands. This thin order book does not include several prominent airlines which have placed options for the 2000. These include American Eagle and Business Express.

Specifications (Saab 2000):

Powerplant: two Allison AE 2100A turboprops, each rated at 3,076 kW (4,125 shp)

Dimensions: length: 27.03 m (88 ft 8.25 in); height: 7.73 m (25 ft 4 in); wing span: 24.76 m (81 ft 2.75 in)

Weights: empty operating: 13,500 kg (29,762 lb); MTOW: 22,000 kg (48,500 lb)

Performance: cruise speed: 653 km/h (353 kts); range: 2,324 km (1,255 nm)

Passengers: 50

Shorts 330/360

The 330 series is flown by military and civilian operators

The last aircraft designed and built by Short Brothers of Northern Ireland, the 330 and 360, are unpressurized twin-turboprop passenger and utility aircraft. Both of the boxy, high-wing designs are powered by Pratt Canada PT6As. The 330 seats 30 passengers while the 360 seats 36.

The 330 programme began in the early 1970s as a derivative of the Shorts Skyvan utility aircraft. The first version, the 330-200, entered service in August 1976. The 360, a stretched 330-200, first flew in June 1981 and entered service in December 1982. Shorts considered plans to stretch the 360 into the 450, but

these were dropped.

Most 330s were bought by military users. The US Air Force and Army National Guard has over 30 330 freighter variants, designated C-23 Sherpa. The 360 was popular with commuter operators until the current generation of pressurized turboprops (DHC-8-100, Saab 340, ATR 42, etc.) arrived in the mid 1980s. Major 360 operators include Business Express, CCAir, and Flagship. The US military is buying some used 360s for conversion to C-23s.

The 330 ended production in 1989, while the 360 lingered on until 1991. Production totaled 179 330s and 164 360s. Most of these are still in service, and Shorts continues to support them. Shorts, now owned by Canada's Bombardier, is still in business as an aircraft subcontractor and missile manufacturer.

Specifications (360F):

Powerplant: two Pratt & Whitney Canada PT6A-67R turboprops, each rated at 1,062 kW (1,424 shp)
Dimensions: length: 21.58 m (70 ft 9.5 in); height: 7.27 m (23 ft 10.25 in); wing span: 22.8 m (74 ft 9.5 in)
Weights: empty operating: 7,870 kg (17,350 lb); MTOW: 12,292 kg (27,100 lb)
Performance: cruise speed: 400 km/h (216 kts); range: 1,178 km (636 nm)
Passengers: 36

Xian Y-7

Xian has built 100 of these An-24 copies

The Y-7 is a copy of the Antonov An-24 built in China by Xian Aircraft Corp. Like the An-24, the Y-7 is a 48/52-seat high-wing twin-turboprop transport. It is also used for freight operations. The Y-7 development programme was rather simple: Xian got hold of an An-24, took it apart, and reverse engineered it. China tried the same thing with a Boeing 707, but predictably, this was a total failure.

The first of three Y-7 prototypes flew in December 1970. Chinese certification was awarded in 1980. The first version, the Y-7, entered service in early 1984. It was followed by the Y-7-100, which has winglets and other modifications. The Y-7-100 has a three-crew flight deck, a great improvement over the Y-7, which, humorously, needed five crew-members.

As of 1995 Xian had built about 100 Y-7s. Production is continuing, and Xian has started the Y7H programme, a derivative closely related to the An-26. The first Y7H flew in 1988. Other Y-7 derivatives include the Y-7-200, with Pratt & Whitney Canada engines and Western avionics. This could enter service in 1997. The Y-7 can be found primarily in China, where the country's airlines operate them on regional routes. China Northern and China Eastern have about ten each. Air China has six. A few have been exported, including three for Lao Aviation.

Specifications (Y-7-100):

Powerplant: two Dongan (DEMC) WJ5A I turboprops, each rated at 2,080 kW (2,790 shp)
Dimensions: length: 24.218 m (79 ft 5.5 in); height: 8.553 m (28 ft 0.75 in); wing span: 29.666 m (97 ft 4 in)
Weights: empty operating: 14,988 kg (33,042 lb); MTOW: 21,800 kg (48,060 lb)
Performance: cruise speed: 423 km/h (228 kts); range: 910 km (491 nm)
Passengers: 52

British Aerospace 125

BAe built 573 125 business jets before the type evolved into the Raytheon-built -800

The 125 is a series of twin-engine medium-size business jets. As with many post-war British aircraft, the 125 has a complicated parentage. It was created as the de Havilland 125, then became the Hawker Siddeley 125. It was later absorbed into British Aerospace. The 125 first flew in 1962. There were numerous early variants, including the 125-1, -1A, -

1B, -2, -3, -3A, etc. Most of these were built in small batches. They were followed by the -400 and -600. The -600 used a stretched fuselage, and could seat up to 14 passengers (6-8 was normal).

All of these versions were powered by Rolls-Royce Viper turbojets. The Royal Air Force uses 125-2s as Dominie T1 navigational trainers and CC1 and CC2 communications planes. The next 125 was the -700, the first to be powered by Garrett (now AlliedSignal) TFE731-3 turbofans. The -700 can also be distinguished by its longer, pointed nose. It first flew in June 1976 and entered service in 1977.

BAe built a total of 573 125s, including 215 125-700s and 72 -600s. Production ended in 1984, but the type evolved into the 125-800, now Raytheon's Hawker 800. Many 125s are still flying.

Specifications (125-700):

Powerplant: two Garrett TFE731-3-1RH turbofans, each rated at 16.46 kN (3,700 lbst)

Dimensions: length: 15.46 m (50 ft 8.5 in); height: 5.36 m (17 ft 7 in); wing span: 14.33 m (47 ft 0 in)

Weights: empty operating: 5,826 kg (12,845 lb); MTOW: 11,566 kg (25,500 lb)

Performance: cruise speed: 723 km/h (390 kts); range: 4,482 km (2,420 nm)

Passengers: 8

Canadair Challenger

The current production model is the 601-3A

The Challenger is a twin turbofan large, long-range business jet built by Bombardier's Canadair division. The first model, the 600, was powered by Lycoming ALF502 engines. It flew in November 1978, but was replaced by the General Electric CF34-powered Challenger 601 in the early 1980s. The current production model, the 601-3A, is available with Extended Range (ER) modifications. Along with the Gulfstream IV and Falcon 900, the Challenger sells in the high end of the bizjet market. Rock stars,

government ministers from oil-rich countries, and American televangelists use planes like these to avoid accidental downgrading to the economy/cattle section of a 747. The price for this peace of mind: about $20 million for a complete Challenger.

In addition to rich companies and individuals, the Challenger is used by militaries for various roles. Canada's Air Force uses three for electronic warfare training, and Germany's Luftwaffe uses the Challenger for ambulance and transport flights.

The 601-3A will be replaced by the improved 604 in early 1996. It will retain the CF34 powerplant, and features airframe and landing gear improvements. In addition, Canadair is planning the even larger and longer-ranged Global Express. This will take over much of the high end of the market.

Specifications (Challenger 601-3A/ER)

Powerplant: two General Electric CF34-3A turbofans, each rated at 40.66 kN (9,140 lbst) with APR (automatic power reserve)
Dimensions: length: 20.85 m (68 ft 5 in); height: 6.3 m (20 ft 8 in); wing span: 19.61 m (64 ft 4 in)
Weights: empty operating: 11,684 kg (25,760 lb); MTOW: 20,230 kg (44,600 lb)
Performance: cruise speed: 851 km/h (459 kts); range: 6,643 km (3,585 nm)
Passengers: 50

Canadair CL-415

A CL-415 takes a drink before continuing fire-fighting

Canadair's CL-415 is the only amphibious Western aircraft designed specifically to fight fires. Its mission: to land in lakes, suck up water, and drop it on a forest fire. It features a four-tank fire-fighting system with foam chemical injection. The CL-415 can also operate from land, using retractable landing gear. It's easy to recognize: it has a flying boat hull and high wings with top-mounted engine nacelles and water float pylons.

The CL-415 is a turboprop version of the CL-215, a piston-powered design built by Canadair between 1969 and 1990. A total of 125 CL-215s were built,

mostly for Canadian provincial governments, France, Greece, and Spain. Some CL-215s are being re-engined with the CL-415's PW123AF engines to become CL-215Ts. CL-215T conversion kits have been ordered by the Spanish and Quebec governments.

The CL-415 was launched in October 1991 by orders from the French and Quebec governments, and made its maiden flight in December 1993. Deliveries began in late April 1994. Canadair is promoting the CL-415 for a number of alternative missions, including surveillance and sea rescue. No orders have been received, but CL-415 production is continuing. In September 1994 fire-plagued Los Angeles leased a CL-215T for testing, and could order some.

Specifications (CL-415, land-based water bomber operations):

Powerplant: two Pratt & Whitney Canada PW123AF turboprops, each rated at 1,775 kW (2,380 shp)
Dimensions: length: 19.82 m (65 ft); height: 8.98 m (29 ft 6 in); wing span: 28.63 m (93 ft 11 in)
Weights: empty operating: 12,333 kg (27,190 lb); MTOW: 19,890 kg (43,850 lb)
Performance: cruise speed: 269 km/h (145 kts); range: 2,428 km (1,310 nm)
Payload: 6,123 kg (13,500 lb)

Canadair Global Express

Built for the Executive who already has everything, the Global Express costs as much as a refurbished 747

The Global Express is a new high-end business jet derivative of the Canadair Challenger designed for very long range (trans-Pacific) flights. It will use new supercritical wings and flies at Mach 0.88 speed. It has the same cabin length as Canadair's RJ airliner.

The Global Express entered the advanced design phase in February 1993. In March, Rolls-Royce/BMW's BR710 was chosen as powerplant (as on Gulfstream's GV), and in September Mitsubishi signed a risk sharing agreement covering up to 20%

of the aircraft's development costs. Mitsubishi is building the wing and centre fuselage. Honeywell is providing the avionics package. On 20 December 1993, Bombardier officially launched the project (it will actually be called the Bombardier Global Express, the first aircraft to bear the corporate name). The company wanted 40 firm orders before launch, but settled for 30, with eight options. The Global Express sells for about $30 million. It competes directly with Gulfstream's GV for orders from very rich people, although for the same price these people could buy a used 747 with a lavish new interior.

The Global Express will make its first flight in September 1996, followed by first deliveries to completion centres in December 1997. Canadian and US certification is scheduled for March 1998.

Specifications (Global Express):

Powerplant: two BMW/Rolls-Royce BR710-48-C2 turbofans, each rated at 65.3 kN (14,690 lbst)
Dimensions: length: 30.3 m (99 ft 5 in); height: 7.47 m (24 ft 6 in); wing span: 28.5 m (93 ft 6.5 in)
Weights: empty operating: 18,460 kg (40,612 lb); MTOW: 41,277 kg (91,000 lb)
Performance: cruise speed: 888 km/h (480 kts); range: 11,723 km (6,330 nm)
Passengers: 8-19

Cessna Caravan

The Cessna Caravan is used by Federal Express and several Brazilian operators

Cessna's 208 Caravan I is a rugged, high-wing unpressurized single turboprop aircraft designed for business and utility transport. A stretched version is known as the 208B Caravan 1B. Its cousin, the Caravan II, is built by France's Reims Aviation as a business aircraft carrying between six and nine passengers.

Cessna began the Caravan programme as a successor to its earlier piston-powered utility aircraft - the Cessna 180, 185, and 206. The Caravan, also called

Cargomaster, first flew in December 1982. Certification was awarded in October 1984, and deliveries began in February 1985. The Caravan is primarily used for small package delivery, although it can also seat up to fourteen passengers. Federal Express is by far the most significant customer. The freight company placed its first Caravan order in December 1983, and has since ordered about 240 aircraft. Most of these are Caravan 1Bs. Brazil's TAM is the second largest Caravan user.

The faster Caravan II first flew in September 1983. Deliveries began in April 1985. Reims builds small numbers of Caravan IIs, and as of late 1995 had delivered fewer than 100.

Cessna continues to build the Caravan, and as of late 1995 the company had delivered over 700 planes.

Specifications (Caravan I)

Powerplant: one Pratt & Whitney Canada PT6A-114 turboprop rated at 447 kW (600 shp)
Dimensions: length: 11.46 m (37 ft 7 in); height: 4.27 m (14 ft 0 in); wing span: 15.88 m (52 ft 1 in)
Weights: empty operating: 1,724 kg (3,800 lb); MTOW: 3,311 kg (7,300 lb)
Performance: cruise speed: 341 km/h (184 kts); range: 2,009 km (1,085 nm)
Passengers: 9

Cessna Citation 500/I/CitationJet

The CitationJet is the first FJ-44 powered jet to enter service

The Citation 500, I, and CitationJet are the entry-level models of the long-running Citation business jet series. Cessna announced its intention to develop a new eight-seat bizjet in October 1968. This became the Citation 500, the first model in the Citation series. The 500 first flew in September 1969. Certification came in February 1972. Production ended in the late 1970s, and the 500 was succeeded by the Citation I.

The Citation I was introduced in September 1976

along with the II and. The Citation I was certified in December 1976. A total of 698 Citation 500s and Is were built by the time production of the latter model ended in 1985.

The CitationJet is the smallest Citation, with seating for six passengers. Also known as the Model 525, the CitationJet is the first business jet to enter service with the Williams/Rolls-Royce FJ44 turbofan, which also powers the Swearingen SJ30. First flight took place in April 1991, and certification was obtained in October 1992, but first deliveries were not until March 1993. By 1996, Cessna had delivered over 100 CitationJets, and production is continuing.

Cessna also built two examples of a tandem-seat military trainer version of the CitationJet, but this failed to win an important US military competition.

Specifications (Citationjet):

Powerplant: two Williams International/Rolls-Royce FJ44 turbofans, each rated at 8.45 kN (1,900 lbst)
Dimensions: length: 12.98 m (42 ft 7.25 in); height: 4.18 m (13 ft 8.5 in); wing span: 14.26 m (46 ft 9.5 in)
Weights: empty operating: 2,823 kg (6,224 lb); MTOW: 4,717 kg (10,400 lb)
Performance: cruise speed: 709 km/h (383 kts); range: 2,696 km (1,456 nm)
Passengers: 6

Cessna Citation II/SII/V

The Citation II business jet entered service in 1978

The Citation II, S/II and V are straight-wing light twinjet designs powered by Pratt & Whitney Canada JT15D turbofans. The first 6/10-seat Citation II flew in January 1977. Certification was received in April 1978. Production of the II ended in 1985 in favour of the S/II, but resumed in 1987. The S/II is an improved 8-10 seat version of the II, first announced in October 1983. Also known as Model S550, the S/II was certified in July 1984. In 1985, the US Navy

procured 15 S/IIs, designated T-47A, for training.

In 1987, Cessna announced development of the Model 560 Citation V. The V is a development of the S/II with a stretched fuselage and more powerful engines. It was certified in December 1988. First deliveries came in April 1989. In September 1993 the V was updated with more powerful engines and new avionics. Now known as the Citation V Ultra, this version was FAA certified and delivered to customers in late June 1994.

The last Citation II was delivered in September 1994. Cessna built a total of 830 IIs and S/IIs. Production of the Citation V Ultra is continuing, and by the end of 1995 Cessna had delivered over 300 Vs and V Ultras. The V Ultra sells for about $5.4 million.

Specifications (Citation II):

Powerplant: two Pratt & Whitney Canada JT15D-4B turbofans, each rated at 11.12 kN (2,500 lbst)
Dimensions: length: 14.29 m (47 ft 2.5 in); height: 4.57 m (15 ft 0 in); wing span: 15.9 m (52 ft 2 in)
Weights: empty operating: 3,504 kg (7,725 lb); MTOW: 6,396 kg (14,100 lb)
Performance: cruise speed: 713 km/h (385 kts); range: 3,260 km (1,760 nm)
Passengers: 6-10

Cessna Citation III/VI/VII

The Citation VII has more powerful engines than its cousins, the III and VI

The Citation III, VI, and VII are swept-wing medium-size bizjets powered by two AlliedSignal TFE731 turbofans. Cessna introduced the III as an entirely new design in September 1976. Also known as the Model 650, the III first flew in May 1979. First production deliveries were in December 1982.

Cessna delivered 214 Citation IIIs, with production ending in 1992. The III was replaced by the Citation

IV, which Cessna introduced in 1989. A development of the III, the IV had a particularly short life span. It was discontinued in May 1990 and replaced with the Citation VI and VII.

The Citation VI is a lower-priced version of the III, with the same engines, cabin size and performance. Deliveries began in mid 1991. Production wound down in 1995 after about 38 had been delivered, and the type will be replaced with the Citation Excel.

The Citation VII is essentially the III/VI with more powerful engines, for hot-and-high operations. Flight tests began in early 1991, with first deliveries in early 1992. Citation VII Production is continuing, and by the end of 1995 Cessna had delivered over 50 VIIs. The VII sells for about $8.6 million. Cessna plans to upgrade the VII sometime in the next few years.

Specifications (Citation VI):

Powerplant: two AlliedSignal/Garrett TFE731-3B-100S turbofans, each rated at 16.24 kN (3,650 lbst)
Dimensions: length: 16.9 m (55 ft 5.5 in); height: 5.12 m (16 ft 9.5 in); wing span: 16.31 m (53 ft 6 in)
Weights: empty operating: 5,851 kg (12,900 lb); MTOW: 9,979 kg (22,000 lb)
Performance: cruise speed: 874 km/h (472 kts); range: 4,345 km (2,346 nm)
Passengers: 6

Cessna Citation X

The Citation X is the fastest business jet available

A trans-continental/trans-Atlantic mid-size business jet, the Citation X is the largest model of the Citation family. It seats up to 12 passengers. Also known as the Cessna 750, the 'X' stands for 'ten', not 'experimental'. Cessna claims that the X will be the 'fastest commercial aircraft in the world apart from Concorde,' with a Mach 0.9 maximum operating speed. This gives the plane a New York to Los Angeles flight time of four hours.

The Citation X is powered by two Allison AE3007C engines mounted on the rear fuselage. It has supercritical sweptback wings. The forward fuselage and cockpit sections are derived from the Citation VI. There is a Honeywell Primus 2000 integrated avionics system.

Cessna introduced the Citation X in 1990. The first of two Citation X prototypes was rolled out on 15 September 1993. The first flight took place in December 1993. FAA certification was awarded in November 1995. First deliveries will take place in April 1996.

Cessna received orders for 10 Xs by early 1991, and plans to produce 24-36 per year. It will compete with Dassault's Falcon 2000 and Raytheon's Hawker 1000. The Citation X costs about $13 million.

Specifications (Citation X):

Powerplant: two Allison AE 3007C turbofans, each rated at 28.47 kN (6,400 lbst)
Dimensions: length: 22.0 m (72 ft 2 in); height: 5.77 m (18 ft 11 in); wing span: 19.48 m (63 ft 11 in)
Weights: empty operating: 9,163 kg (20,200 lb); MTOW: 15,649 kg (34,500 lb)
Performance: cruise speed: 927 km/h (500 kts); range: 6,117 km (3,300 nm)
Passengers: 12

Cessna Citation Bravo/Excel

The Excel and Bravo use the new PW500 engines

The Bravo and Excel are the two newest models in Cessna's Citation business jet line. Both are straight-wing light twinjet designs powered by Pratt & Whitney Canada's PW500 series turbofan. Cessna announced the Bravo at the September 1994 Farnborough Air Show. The new six-seat Bravo will replace the Citation II in the product line. The new aircraft will be based on the II's airframe, but use

PW530 engines instead of JT15Ds.

The Bravo made its first flight in April 1995 and will be certified in April 1996. Deliveries will begin in June 1996. The Bravo costs $4.6 million in 1996 dollars.

In October 1994 Cessna unveiled its Citation Excel. The new 7-8 seat model will use the Citation V wing and a widebody fuselage derived from the Citation. It will be the first light business jet with a stand-up cabin. The Excel will be powered by PW545A engines, and will sell for about $7 million. The first Excel will fly in March 1996, with FAA certification scheduled for summer 1997. Deliveries will begin in December 1997, and in 1995 Cessna claimed to hold orders for 70 Excels. The type will replace the Citation VI in the product line.

Specifications (Citation Bravo):

Powerplant: two Pratt & Whitney Canada PW530A turbofans, each rated at 12.23 kN (2,750 lbst)

Dimensions: length: 14.29 m (47 ft 2.5 in); height: 4.57 m (15 ft 0 in); wing span: 15.9 m (52 ft 2 in)

Weights: empty operating: 3,802 kg (8,383 lb); MTOW: 6,486 kg (14,300 lb)

Performance: cruise speed: 730 km/h (394 kts); range: 3,685 km (1,990 nm)

Passengers: 6-10

Dassault Falcon 10/20/100/200

Many early Dassault business jets like this Falcon 20 found military roles as well

The early Dassault Mystere (now Falcon) series was a line of twinjet business aircraft. It began as the Mystere 20, an 8-10 seat model first flown in 1963. It used General Electric CF700 engines. Used primarily as an executive transport, the 20 also found extensive use as a military trainer and freighter.

The 20 was followed by the Falcon 200, which used Garrett (now AlliedSignal) ATF3-6 engines. Many of these are used for maritime surveillance. The US

Coast Guard uses the type as the HU-25A Guardian, while the French Navy uses it as the Gardian.

The Falcon 10 was the next model. A scaled-down 4-7 seat version of the 20 powered by Garrett TFE731 engines, the first of three Falcon 10 prototypes flew in December 1970. Deliveries began in 1973. The 10 was replaced by the 100, a higher take-off weight version with one extra window on the starboard side.

A total of 226 10/100s was built, with the last example delivered in 1990. This was followed by the last of 514 20/200s in 1991. Many are still in service, and some 60 Falcon 20s have now been retrofitted with TFE731 engines. Dassault have gone on to build the Falcon 2000 as a replacement aircraft.

Specifications (Falcon 100):

Powerplant: two Garrett TFE731-2 turbofans, each rated at 14.4 kN (3,230 lbst)

Dimensions: length: 13.86 m (45 ft 5.75 in); height: 4.61 m (15 ft 1.5 in); wing span: 13.08 m (42 ft 11 in)

Weights: empty operating: 5,055 kg (11,145 lb); MTOW: 8,755 kg (19,300 lb)

Performance: cruise speed: 912 km/h (492 kts); range: 2,900 km (1,565 nm)

Passengers: 8

Dassault Falcon 50/900

LIke the Falcon 50, the 900 is recognisable by its distinctive trijet configuration

The Falcon 50 and 900 are a family of three-turbofan long-range business jets produced by France's Dassault. Both are powered by AlliedSignal/Garrett TFE731 turbofans. The Falcon 50, Dassault's first trijet, was first flown in November 1976. Deliveries began in July 1979. In addition to business users, the Falcon 50 is operated by numerous governments for rescue, VIP, medical, and other duties.

The Falcon 900 is derived from the 50. It features greater, intercontinental range and a wider, longer fuselage than the 50. It can accommodate up to 19

passengers. The 900 made its first flight in September 1984 and deliveries began in December 1986. Like the 50, the 900 is available for non-business applications. Japan's Maritime Safety Agency operates two 900s for long-range ocean surveillance.

The current production model is the Falcon 900B. Introduced in 1991, the 900B features uprated, more efficient engines. In October 1994 Dassault also announced a long-range variant called the 900EX. It was rolled out in March 1995, with certification expected in March 1996. It will have new Honeywell Primus 2000 avionics and a range of 8,334 km (4,500 nm). As of late 1995 Dassault has delivered over 240 Falcon 50s and over 150 Falcon 900s. Both types are still available, but production of the 50 is winding down.

Specifications (Falcon 900B):

Powerplant: three AlliedSignal TFE731-5BR-1C turbofans, each rated at 21.13 kN (4,750 lbst)

Dimensions: length: 20.21 m (66 ft 3.75 in); height: 7.55 m (24 ft 9.25 in); wing span: 19.33 m (63 ft 5 in)

Weights: empty operating: 10,240 kg (22,575 lb); MTOW: 20,640 kg (45,500 lb)

Performance: cruise speed: 927 km/h (500 kts); range: 7,408 km (3,900 nm)

Passengers: 19

Dassault Falcon 2000

The Falcon 2000 can make intercontinental flights

The Falcon 2000 is an eight-passenger twinjet business aircraft with intercontinental range. Designed by Dassault as a replacement for the Falcon 20/200, the 2000 competes with Raytheon's Hawker 1000. It uses the same fuselage cross section as the Falcon 900, but is about one metre (3.2 feet) shorter.

Dassault began market studies for the 2000 in 1987. At the 1989 Paris Air Show they revealed details of the plane, originally called Falcon X. Design

work was performed with the Dassault CATIA computer design system.

The Falcon 2000 made its first flight in March 1993. It received European JAA certification in December 1994. FAA certification and first deliveries came in February 1995. The Falcon 2000 sells for about $16 million. Although designed and built by Dassault in France, the Falcon 2000 makes extensive use of foreign subcomponents. It is the only application so far for the CFE738 turbofan, a joint creation of General Electric and AlliedSignal/Garrett. Italy's Alenia is a programme partner, building the aft fuselage section and engine nacelles. Beyond the baseline 2000, Dassault is considering a stretch version for regional airline operations, similar to the Canadair RJ.

Specifications (Falcon 2000):

Powerplant: two General Electric/AlliedSignal CFE738 turbofans, each rated at 26.7 kN (6,000 lbst)
Dimensions: length: 20.23 m (66 ft 4.5 in); height: 6.98 m (22 ft 10.75 in); wing span: 19.33 m (63 ft 5 in)
Weights: empty operating: 8,936 kg (19,700 lb); MTOW: 15,875 kg (35,000 lb)
Performance: cruise speed: 685 km/h (370 kts); range: 5,556 km (3,000 nm)
Passengers: 10

Gulfstream II/III/IV

Developed from the GII and III, the GIV is the current production Gulfstream business jet

A series of large, long-ranged twin engine business jets, the Gulfstream family is currently in production as the GIV. It is a close relative of the earlier GII and GIII. The first Gulfstream jet, the GII, made its first flight in 1966. The GIII, featuring a longer fuselage and winglets, flew in December 1979. A total of 462 IIs and IIIs were built, with production of the III ending in 1988.

Both planes used Rolls-Royce Spey engines.

Incidentally, there was a GI, an unrelated propeller transport built by Grumman in the 1960s.

The GIV programme began in 1982. It features digital avionics, a redesigned wing, and a stretched fuselage seating 19 passengers. It is powered by Rolls-Royce Tay turbofans. The GIV first flew in September 1985, with first deliveries in late 1986. Selling for approximately $24 million, the GIV was the most expensive business jet on the market until Gulfstream launched its GV and Canadair its Global.

As of late 1995 Gulfstream had delivered over 270 GIVs, and production is continuing. In US military service, the GII/III/IV is designated C-20. A C-20 served as General Norman Schwarzkopf's airborne command post in the war with Iraq. Gulfstream is also marketing a reconnaissance version, the SRA-4.

Specifications (Gulfstream IV):

Powerplant: two Rolls-Royce Tay Mk 611-8 turbofans, each rated at 61.6 kN (13,850 lbst)

Dimensions: length: 26.92 m (88 ft 4 in); height: 7.45 m (24 ft 5.125 in); wing span: 23.72 m (77 ft 10 in)

Weights: empty operating: 19,278 kg (42,500 lb); MTOW: 33,203 kg (73,200 lb)

Performance: cruise speed: 936 km/h (505 kts); range: 6,728 km (3,633 nm)

Passengers: 19

Gulfstream V

The GV competes head-to-head with Canadair's Global Express

The GV will be a longer GIV with new wings and engines. With eight passengers it will have trans-Pacific range, flying up to 12,038 km (6,500 nm) at Mach 0.8. The GV can seat up to 19 passengers on shorter routes. Gulfstream first announced the GV in October 1991. In September 1992 Gulfstream launched its GV with a $250 million investment from chief financial backer Forstmann Little. This amount should cover total research and development costs.

Also in 1992, GV supplier contracts were awarded. In September the new Rolls-Royce/BMW BR710 was selected for the GV, launching the BR700 turbofan project. At the Paris Air Show Northrop Grumman (Vought, at the time) was tapped to build the GV's wings. Honeywell is providing its SPZ-8000 avionics package, and Gulfstream is offering a head-up display (HUD) as an option.

The GV was rolled out in September 1995. Certification is scheduled for October 1996. Seagram Co. will be the first customer to get a GV in November 1996.

The GV sells for $30-35 million. It competes directly with Canadair's Global Express. According to Gulfstream, firm GV sales are up to 55-60 as of late 1995.

Specifications (Gulfstream V):

Powerplant: two BMW/Rolls-Royce BR710-48 turbofans, each rated at 65.61 kN (14,750 lbst)

Dimensions: length: 29.39 m (96 ft 5 in); height: 7.72 m (25 ft 4 in); wing span: 28.5 m (93 ft 6 in)

Weights: empty operating: 21,228 kg (46,800 lb); MTOW: 40,370 kg (89,000 lb)

Performance: cruise speed: 850 km/h (459 kts); range: 12,038 km (6,500 nm)

Passengers: 8-19

Israel Aircraft Industries Astra

The Astra is now built as the SP with a new wing design

The Astra, Israel's only jet aircraft now in production, is a 6/9-seat twin engine business jet designed for transcontinental operations. It first flew in March 1984, and deliveries began in June 1986. The Astra was derived from IAI's first bizjet, the Westwind, which was designed by Rockwell International as the Jet Commander. IAI built a total of 248 Westwinds, with production ending in 1988.

The Astra looks like the Westwind, but only the engine nacelles and tail unit remain the same. The

Astra fuselage is bigger, and there are new sweptback wings. The Astra sells for about $7 million. Since Astra number 42, IAI has been building the Astra SP. The SP features a new interior, range improvements, and a new EFIS. Astra SP deliveries began in 1991.

For the future, IAI is working with Russia's Yakovlev Design Bureau to build a larger, transatlantic business jet. Known as the Galaxy, the concept was originally called the Astra IV. It uses the Astra SP wing, with a new fuselage capable of seating 8-10 passengers, or 19 in a shuttle version. While Yakovlev was originally going to build this fuselage, IAI may look for a different partner.

The Galaxy will use two Pratt & Whitney Canada PW305 turbofans and will sell for about $13 million. The Galaxy could enter service in 1997.

Specifications (Astra SP):

Powerplant: two AlliedSignal TFE731-3C-200G turbofans, each rated at 16.46 kN (3,700 lbst)
Dimensions: length: 16.94 m (55 ft 7 in); height: 5.54 m (18 ft 2 in); wing span: 16.05 m (52 ft 8 in)
Weights: empty operating: 5,999 kg (13,225 lb); MTOW: 10,659 kg (23,500 lb)
Performance: cruise speed: 858 km/h (463 kts); range: 5,211 km (2,814 nm)
Passengers: 6

Learjet 23/24/25/28/29

The Model 23, the first Learjet first flew in 1963

The first Learjets were the Model 23/24/25, a series of six-seat business jets powered by General Electric CJ610 turbojets. Designed by William Lear, the first Learjets used wingtip fuel tanks. They were the first business jets produced in Wichita, Kansas, now a world centre of business jet manufacture.

The Model 23 prototype flew in October 1963. Deliveries began one year later. The Model 23 was followed in 1966 by the Model 24 and its

subvariants, the 24B, C, D, etc. These featured uprated engines, tail unit modifications, and other improvements. The Model 25, a stretched Model 24 with room for eight passengers, first flew in August 1966. The first production version was the Model 25D, first delivered in October 1967. The final Model 25 versions were the long-range Model 25F and G. Production of the 25F began in 1970.

The Model 28 and 29 Longhorn were developments of the Model 25 with much wider wings and winglets instead of fuel tanks. They offered superior performance, especially for take-off and landing. They received FAA certification in January 1979, with first deliveries just after. The Learjet Twenty series ended production in 1985. Learjet built a total of 741 Model 23-29s, and many are still in service.

Specifications (Learjet 24F):

Powerplant: two General Electric CJ610-8A turbojets, each rated at 13.1 kN (2,950 lbst)

Dimensions: length: 13.18 m (43 ft 3 in); height: 3.73 m (12 ft 3 in); wing span: 10.84 m (35 ft 7 in)

Weights: empty operating: 3,234 kg (7,130 lb); MTOW: 6,123 kg (13,500 lb)

Performance: cruise speed: 774 km/h (418 kts); range: 2,512 km (1,355 nm)

Passengers: 6

The Model 35 was the first to use the TFE731 turbofan

The first Learjet to use the Garrett (now AlliedSignal) TFE731 turbofan was the Model 26, a re-engined Model 25 which flew in January 1973. It became the Model 35 and 36, which were certified in July 1974. The two models are similar 6/8-seat mid-sized designs, but the 36 carries additional internal fuel and is capable of intercontinental range. The 35 is transcontinental.

The US Air Force operates approximately 80 Model

35As, this designated C-21A. They are used for high-priority equipment transport, passenger transport, and other missions.

Learjet built a total of 673 Model 35s and 62 Model 36s. Production of both types ended in the early 1990s. Learjet introduced the Model 31 in September 1987. It combines the fuselage and powerplant of the Model 35A/36A with the wing of the 55. A total of 25 were delivered from late 1988 until the end of 1990, and an additional 11 in 1991. It was replaced by the Model 31A, which was FAA certified in July 1991. The 31A features an improved avionics package and a heated windshield. Production of the Model 31 continues, and as of late 1995 Learjet had delivered over 100 aircraft. The Model 31 will be replaced by the Model 45.

Specifications (35/36A):

Powerplant: two AlliedSignal/Garrett TFE731-2-2B turbofans, each rated at 15.6 kN (3,500 lbst)

Dimensions: length: 14.83 m (48 ft 8 in); height: 3.73 m (12 ft 3 in); wing span: 12.04 m (39 ft 6 in)

Weights: empty operating: 4,590 kg (10,119 lb); MTOW: 8,300 kg (18,300 lb)

Performance: cruise speed: 852 km/h (460 kts); range: (36A) 4,671 km (2,522 nm)

Passengers: 6

Learjet 55/60

The intercontinental Learjet 60 replaced the Model 55

The Model 55 is a medium-range 4-8 seat Learjet with a new wing and a stand-up cabin. It is a descendant of the Model 28. A prototype 55 flew in April 1979. Certification was obtained in March 1981, with customer deliveries in April.

The final production version of the 55 was the 55C, certified in December 1988. It incorporated ventral Delta-fins to improve performance. The 55C/ER is an extended range variant with an additional fuel

tank. The 55C/LR is a long-range variant with another fuel tank. The Model 55 has been replaced by the Model 60, and the last of 147 Model 55s was delivered in 1991.

Learjet introduced the transcontinental Model 60 in October 1990. It is based on the Model 55C but is 1.1 m (43 in) longer, with greater flexibility for interior design and more baggage space. It also has new Pratt & Whitney Canada PW305 turbofans and a Rockwell Collins Pro Line 4 avionics package. A Model 55 modified to resemble a Model 60 made its first flight in October 1990. On 15 June 1992 the first production Model 60 made its first flight. It received its FAA certification in January 1993. First delivery, to furniture maker Herman Miller, came the same month.

Specifications (Learjet 60):

Powerplant: two Pratt & Whitney Canada PW305 turbofans, each rated at 20.46 kN (4,600 lbst)
Dimensions: length: 17.88 m (58 ft 8 in); height: 4.47 m (14 ft 8 in); wing span: 13.34 m (43 ft 9 in)
Weights: empty operating: 6,278 kg (13,840 lb); MTOW: 10,319 kg (22,750 lb)
Performance: cruise speed: 858 km/h (463 kts); range: 5,074 km (2,740 nm)
Passengers: 9

The Learjet 45 was the first to benefit from the Bombardier empire

Learjet unveiled its latest project, the Model 45, in September 1992. An entry/mid-level bizjet, the 45 sells for $6 million and seats 8-10 passengers in a stand-up cabin. It has new wings (with winglets) and a new fuselage and tail unit. It will replace the Model 31. The Model 45 is powered by two AlliedSignal

TFE731-20 turbofans. It has a Honeywell Primus 1000 avionics package with EFIS displays and Primus 650 weather radar.

Learjet has awarded airframe subcontracts to Short Brothers of Northern Ireland (fuselage and tail unit) and de Havilland Canada (wings). Like Learjet, these companies are part of the Bombardier empire, making the Model 45 the first pan-Bombardier bizjet.

The Model 45 was rolled out in September 1995 and made its first flight in October. There are five aircraft in the flight test programme. The Model 45 will be FAA certified in December 1996 with first deliveries in early 1997. Learjet claims that it has sold over 100 Model 45s already, and says it will build 30 in 1997. Four have been ordered by Singapore Airlines for use as aircrew trainers.

Specifications (Learjet 45):

Powerplant: two AlliedSignal TFE731-20 turbofans, each rated at 15.57 kN (3,500 lbst)
Dimensions: length: 17.89 m (58 ft 8.5 in); height: 4.48 m (14 ft 8.5 in); wing span: 13.35 m (43 ft 9.5 in)
Weights: empty operating: 5,307 kg (11,700 lb); MTOW: 8,845 kg (19,500 lb)
Performance: cruise speed: 859 km/h (464 kts); range: 4,074 km (2,200 nm)
Passengers: 8-10

Mitsubishi MU-2

Mitsubishi built 755 MU-2s, production ending in 1986

Mitsubishi's MU-2 is a high-wing twin-turboprop utility transport built for a variety of applications. It seats up to 11 passengers and is distinguished by wingtip fuel tanks. Its landing gear retract into the fuselage. The MU-2 first flew in September 1963. The first version was the MU-2A, a series of three prototypes. The MU-2B was the first production version, followed by the MU-2C, D, EP, and other

variants. The MU-2G and J have a stretched fuselage. The final version was the MU-2P.

The Japanese army uses the MU-2C as the LR-1 liaison and reconnaissance aircraft, and the air force uses the MU-2E for search and rescue. Most MU-2s were built in Japan, but some were assembled from Mitsubishi-supplied kits at a plant in Texas. Japanese-assembled MU-2s are called Marquise. MU-2s completed in Texas are called Solitaire.

MU-2 production ended in March 1986 after 755 were built. Most went to export customers, particularly in North America. Many are still flying today, but the aircraft has acquired a slight reputation problem. About 20% of the fleet have been involved in accidents. Also, as a fast, long-range aircraft, it is popular for Caribbean drug-smuggling operations.

Specifications (MU-2N):

Powerplant: two Garrett TPE 331-5-252M turboprops, each rated at 579 kW (776 ehp)
Dimensions: length: 12.02 m (39 ft 5 in); height: 4.17 m (13 ft 8 in); wing span: 11.94 m (39 ft 2 in)
Weights: empty operating: 3,205 kg (7,065 lb); MTOW: 5,250 kg (11,575 lb)
Performance: cruise speed: 550 km/h (295 kts); range: 2,330 km (1,260 nm)
Passengers: 11

Piaggio P.180 Avanti

Sadly, the spectacular P.180 has not sold well

If upmarket high-tech mail order companies sold planes, this is what they would sell. Piaggio's P.180 looks like something children would draw if they were trying to create the most novel and complicated aircraft imaginable. A 6-10 seat corporate transport/utility aircraft, the P.180 features foreplanes, a main wing located on the rear fuselage, and twin turboprops mounted in a rear-facing pusher configuration. Composite parts make up 17% of the aircraft by empty weight.

Piaggio began the P.180 programme in 1979. It first flew in September 1986. Italian certification came in March 1990, followed by FAA certification in May. The first production P.180 was delivered in October 1990. While designed and built in Italy, much of the P.180 is built in the US, including the fuselage. Rockwell's Collins unit provides most of the avionics, including a digital autopilot and a three-tube electronic flight instrumentation system (EFIS).

Unfortunately, weird, imaginative planes do not always fit the corporate image. The competing Beech Starship is now out of production, and Piaggio builds fewer than 12 P.180s per year. Strangely, the US Army National Guard ordered four. Even stranger, in 1992 the P.180 became the first Western business plane sold to Bulgaria.

Specifications (P.180):

Powerplant: two Pratt & Whitney Canada PT6A-66 turboprops, each rated at 1,107 kW (1,485 shp)
Dimensions: length: 14.41 m (47 ft 4 in); height: 3.94 m (12 ft 11 in); wing span: 14.03 m (46 ft 1 in)
Weights: empty operating: 3,402 kg (7,500 lb); MTOW: 5,239 kg (11,550 lb)
Performance: cruise speed: 482 km/h (260 kts); range: 2,594 km (1,400 nm)
Passengers: 6-10

Pilatus PC-12

The PC-12 is the latest multi-role aircraft from Pilatus

The PC-12 is the latest in a line of single-turboprop utility aircraft developed by Swiss aircraft manufacturer Pilatus. The company developed the PC-12 mainly for the small-package delivery market. The aircraft is pressurized and certified for single-pilot operations. Development work began in the mid 1980s. Pilatus unveiled a mockup in October 1989. The PC-12 made its first flight in May 1991. Swiss certification was awarded in April 1994, followed by first deliveries in May. The first aircraft went to launch customer Zimex Aviation. US FAA certification was awarded in July 1994.

The PC-12 is marketed in five versions: executive,

corporate commuter (PC-12P), freighter (PC-12F), military, and combi. The executive model seats six passengers; the commuter version can accommodate nine passengers; the combi seats four passengers and cargo.

The passenger version features all-digital avionics. The military PC-12 can perform a variety of missions including medevac, patrol, forward air control, training and paradrop. Several PC-12s have been delivered in ambulance configuration to Australia's Royal Flying Doctor Service. While Pilatus can build up to 56 PC-12s per year, it has only received firm orders for several dozen aircraft. The PC-12 competes with the Cessna Caravan in the freight market and with Socata's TBM 700 in other utility and passenger markets.

Specifications (PC-12):

Powerplant: Pratt & Whitney Canada PT6A-67B turboprop, each flat-rated at 895 kW (1,200 shp) for take-off
Dimensions: length: 14.4 m (47 ft 2 in); height: 4.26 m (14 ft); wing span: 16.08 m (52 ft 9 in)
Weights: empty operating: 2,386 kg (5,260 lb); MTOW: 4,000 kg (8,818 lb)
Performance: cruise speed: 496 km/h (268 kts); range: 2,965 km (1,600 nm)
Passengers: 9

Pilatus Britten-Norman Islander

The Islander is a veritable workhorse, able to carry freight, passengers or sensors

Looking for a cheap way to fly 10 people to the middle of a crater? Looking for drug-smuggling boats? Looking for locusts in Africa? If the answer to any of these is yes, you should consider the Islander, a cheap, rugged twin-engine utility aircraft which can be used in any role. Britten-Norman (now part of Switzerland's Pilatus) began designing the Islander in April 1964. A prototype flew in June 1965. The first

production BN-2 began operations in August 1967.

The Islander is available with piston or turboprop engines. The latter aircraft, the BN-2T Turbine Islander, entered service in late 1981. Military Islanders are called Defenders. An 18-seat stretch version of the piston Islander, with a third engine mounted on the tailplane, is known as the Trislander. Some 91 Trislanders were built between 1971 and 1984. With its large nose and high wings, the Islander makes an ideal sensor platform for military and civil duties. These can search for a variety of ground and sea targets. One version, the Defender 4000, can carry an air search radar.

Over 1,200 Islanders have been built, and most of these are still in service. Production is continuing at the assembly line on the Isle of Wight.

Specifications (Turbine Islander):

Powerplant: two Allison Engines Model 250-B17C turboprops, each flat-rated at 238.5 kW (320 shp)
Dimensions: length: 10.86 m (35 ft 8 in); height: 4.18 m (13 ft 9 in); wing span: 14.94 m (49 ft)
Weights: empty operating: 1,832 kg (4,040 lb); MTOW: 3,175 kg (7,000 lb)
Performance: cruise speed: 285 km/h (143 kts); range: 1,349 km (728 nm)
Passengers: 8

Raytheon Beechjet

Raytheon's first business jet, the Beechjet was originally designed by Mitsubishi

The Beechjet 400 is a twin turbofan 8-seat entry-level business jet. Built by Raytheon, the Beechjet was designed by Japan's Mitsubishi as the Diamond II. The Diamond first flew in August 1978. Raytheon's Beech unit (now called Raytheon Aircraft Co.) acquired the rights to this plane in late 1985, and rolled out the first Beech-built aircraft in May 1986. Beech moved the production line to Kansas in 1989.

Beech then developed the plane into the Beechjet 400A, first announced in 1989. This version, still in production, features improved performance and greater weights, and new all-digital Rockwell Collins Pro Line 4 avionics. Deliveries of the 400A began in 1990.

In March 1990, the US Air Force selected the Beechjet 400T for its tanker and transport training requirement. This contract was the business jet equivalent of winning the lottery. The service is procuring 180 Beechjets, designated T-1A Jayhawk. Deliveries of these began in January 1992. Japan's Air Self Defence Force has also ordered the 400T for training. The Beechjet sells for just under $5 million. As of late 1995, Raytheon had built about 300 Beechjets, including over 120 T-1As for the USAF.

Specifications (Beechjet):

Powerplant: two Pratt & Whitney Canada JT15D-5 turbofans, each rated at 13.19 kN (2,965 lbst)
Dimensions: length: 13.15 m (43 ft 2 in); height: 4.24 m (13 ft 11 in); wing span: 13.25 m (43 ft 6 in)
Weights: empty operating: 4,833 kg (10,655 lb); MTOW: 7,303 kg (16,100 lb)
Performance: cruise speed: 726 km/h (392 kts); range: 3,343 km (1,805 nm)
Passengers: 8

Raytheon Hawker 800

The Hawker 800 was designed by British Aerospace

The Hawker 800 is a 6/14-seat twin-engine medium-sized business jet with transcontinental range. Originally, it was called the British Aerospace 125-800, the last in a long line of BAe 125 series corporate jets. Raytheon purchased the production line in June 1993, reviving the venerable Hawker name. BAe began the 125-800 programme in 1977, as a development of the 125-700. The 800 uses the TFE731-5 turbofan, an upgrade of the TFE731-3 on

the 125-700, and Rockwell Collins Pro Line II avionics. The first 125-800 flew in May 1983. FAA certification was awarded in December 1984, with first deliveries the same year.

While primarily used by corporate operators, the 800 has found numerous government and military applications. Japan's military is buying 27 for search and rescue, plus three more for flight inspection. The Royal Saudi Air Force has 12 for VIP transport. The US Air Force uses six for flight inspection and navigation, designated C-29A. A Hawker 800 costs about $10 million.

As of late 1995, BAe/Raytheon had built over 250 800s. Production is continuing, and the final assembly line will be moved from Britain to the US in 1997.

Specifications (Hawker 800):

Powerplant: two AlliedSignal TFE731-5R-1H turbofans, each rated at 19.13 kN (4,300 lbst)
Dimensions: length: 15.6 m (51 ft 2 in); height: 5.36 m (17 ft 7 in); wing span: 15.66 m (51 ft 4.5 in)
Weights: empty operating: 7,076 kg (15,600 lb); MTOW: 12,428 kg (27,400 lb)
Performance: cruise speed: 741 km/h (400 kts); range: 4,778 km (2,580 nm)
Passengers: 8

Raytheon Hawker 1000

Raytheon are increasing the range of the Hawker 1000

The Hawker 800 is a 8-15 seat twin-engine medium-sized business jet with transcontinental to intercontinental range. It was originally designed by British Aerospace as a stretched and modified 125-800 (see Hawker 800). Raytheon purchased the production line in June 1993.

In addition to the 0.8 m (2 ft 9 in) stretch, the 1000 differs from the 800 in several ways. The 1000 has new engines, new avionics, and interior upgrades. However, the new design is only barely intercontinental, and as of late 1995 Raytheon was pondering airframe modifications necessary to

improve this.

BAe began work on the 1000 in early 1988. It made its first flight in June 1990. British CAA and US FAA certification came in October 1991.

So far, the 1000 has gone mostly to corporate operators. Pratt & Whitney Canada and parent company United Technologies have four - unsurprising, since the 1000 uses Pratt & Whitney Canada PW305 engines. Aravco Limited and J.C. Bamford Excavators have two each. The 1000 sells for about $12-13 million and competes directly with Dassault's Falcon 2000. About 50 1000s were built by late 1995, but deliveries have been temporarily halted while the aircraft design is modified. The final assembly line will be moved from Britain to the US in 1997.

Specifications (Hawker 1000):

Powerplant: two Pratt & Whitney Canada PW 305B turbofans, each rated at 23.13 kN (5,200 lbst)
Dimensions: length: 16.42 m (53 ft 10.5 in); height: 5.21 m (17 ft 1 in); wing span: 15.66 m (51 ft 4.5 in)
Weights: empty operating: 7,811 kg (17,220 lb); MTOW: 14,060 kg (31,000 lb)
Performance: cruise speed: 867 km/h (468 kts); range: 5,750 km (3,105 nm)
Passengers: 15

Raytheon King Air

The Super King Air 350 is the largest and most expensive King Air turboprop

The Beech (now Raytheon) King Air series comprises a family of pressurized twin-turboprop business and utility aircraft. Used for a variety of civil and military applications, the King Air has been in production for over 30 years. The King Air was developed in the early 1960s, with a first flight in 1964. The first model was the King Air Model 90. This and all subsequent models use two Pratt & Whitney Canada

PT6A engines, except for the King Air B100, which had Garrett TPE331s. Today, there are five main production models. The smallest is the King Air C90SE, followed by the C90B. In the middle are the Super King Air B200 and 300. The top-of-the line model is the Super King Air 350. The King Air is primarily a business transport, and can seat 9-16 passengers. Some are used as commuter airliners, and Beech built 11 Beech 1300s as dedicated 13-seat airliner variants of the B200. Mesa Airlines bought ten of these.

The US Army uses the King Air, designated C-12, for numerous applications. The RC-12 carries a variety of electronic sensors for intelligence missions. As of late 1995 Beech has built over 4,800 King Airs, and production is continuing.

Specifications (King Air C90B):

Powerplant: two Pratt & Whitney Canada PT6A-21 turboprops, each rated at 410 kW (550 shp)
Dimensions: length: 10.82 m (35 ft 6 in); height: 4.34 m (14 ft 3 in); wing span: 15.32 m (50 ft 3 in)
Weights: empty operating: 3,028 kg (6,675 lb); MTOW: 4,581 kg (10,100 lb)
Performance: cruise speed: 457 km/h (247 kts); range: 1,728 km (933 nm)
Passengers: 10

Raytheon Starship

The futuristic Starship has not attracted customers

The Beech (now Raytheon) Starship is a twin-turboprop business aircraft. It is notable for its unusual, high-tech design and exceptionally short life. First, the design: the swept main wings, mounted on the aft fuselage, end with vertical stabilizers, or 'tipsails'. There are two pusher propellers on the back of the main wings. There are smaller variable-geometry wings mounted on the forward fuselage. The pressurized fuselage seats 8-10 passengers. The entire aircraft is composed of composite materials, such as graphite-carbon epoxy and Nomex-honeycomb. This was all very unusual, especially for a

plane built in Wichita, Kansas.

The first Starship was an 85% scale prototype built by the legendary Burt Rutan company, Scaled Composites. This flew in 1983, followed by a 100% prototype in February 1986. The first production Starship flew in April 1989, and received FAA certification in December. In October 1991 Beech introduced an improved model, the Starship 2000A. This version has a greater fuel capacity and interior improvements. It was FAA certified in April 1992.

Like the Piaggio Avanti, the Starship met a very unenthusiastic market. Unlike Piaggio, Beech decided to kill the programme. In 1994 Beech built its 53rd and last Starship, making it the airborne equivalent of the DeLorean sports car. Don't worry, though - the company still has plenty on its airfield, waiting for a customer.

Specifications (Starship):

Powerplant: two Pratt & Whitney Canada PT6A-67A turboprops, each rated at 895 kW (1,200 shp)
Dimensions: length: 14.05 m (46 ft 1 in); height: 3.94 m (12 ft 11 in); wing span: 16.6 m (54 ft 4.75 in)
Weights: empty operating: 4,590 kg (10,120 lb); MTOW: 6,758 kg (14,900 lb)
Performance: cruise speed: 589 km/h (318 kts); range: 2,804 km (1,514 nm)
Passengers: 6

Socata TBM 700

The versatile TBM700 has digital avionics

The TBM 700 is a single turboprop business and utility aircraft. A pressurized low-wing design, the TBM 700 is capable of carrying 6-8 passengers or freight. It can also perform medical evacuation, target towing, and photography missions. It features advanced digital avionics and retractable landing gear. Launched in June 1987, the TBM 700 was created by a joint venture between France's Socata and the USA's

Mooney Aircraft. The first of three prototypes flew in July 1988. The aircraft received French DGAC certification in January 1990, followed by US FAA certification in August and first deliveries in October.

Original plans called for Mooney to build the TBM 700 mid/aft fuselage and wings, but Mooney dropped out of its 30% share of the venture for financial reasons in May 1991. Socata, supported by parent company Aerospatiale, pressed on. For the future, Socata is considering a stretched 8/10-seat model, known as the TBM 700S. Socata builds the TBM 700 at a facility in Tarbes, France. As of late 1995 over 80 aircraft had been delivered. Users so far include various companies and the French Air Force, which ordered six aircraft for liaison. The TBM 700 sells for about $1.5 million.

Specifications (TBM 700):

Powerplant: one Pratt & Whitney Canada PT6A-64 turboprop, rated at 522 kW (700 shp)

Dimensions: length: 10.43 m (34 ft 2.5 in); height: 3.99 m (13 ft 1 in); wing span: 12.16 m (39 ft 10.75 in)

Weights: empty operating: 1,826 kg (4,025 lb); MTOW: 2,984 kg (6,578 lb)

Performance: cruise speed: 555 km/h (300 kts); range: 2,982 km (1,610 nm)

Passengers: 7

Swearingen SJ30

The SJ30 was the first FJ44-powered business jet

The idea behind the SJ30 - a low-cost efficient entry-level business jet - can be traced back to the early 1980s. Noted aircraft designer Ed Swearingen began creating the appropriate airframe. Unfortunately, there were no light, efficient turbofans in the necessary class. But in 1986 Williams International, with later help from Rolls-Royce, designed the FJ44, derived from a cruise missile engine. The new FJ44-powered aircraft design became the SA-30 Fanjet. The next problem was finding a home. For a while, it was

a Gulfstream project, but this did not last. In the late 1980s Jaffe Group provided some funding and the aircraft became the SJ30. A prototype flew in February 1991. In late 1991 Swearingen announced that the SJ30 would be built in Delaware, but this fell through.

Current plans call for the SJ30 to be built in Martinsburg, West Virginia, with help from Taiwanese investors and subcontractors. Swearingen hopes to fly a second prototype in 1995, and claims to have over 60 orders. While the SJ30 is a very innovative design, it is no longer the first in its class. Cessna's CitationJet uses the same engines. It was developed after the SJ30 but entered service in 1992. When SJ30 deliveries begin, hopefully in 1997, there will be an established competitor.

Specifications (SJ30):

Powerplant: two Williams International/Rolls-Royce FJ44 turbofans, each rated at 8.45 kN (1,900 lbst)
Dimensions: length: 12.98 m (42 ft 7 in); height: 4.24 m (13 ft 11 in); wing span: 11.1 m (36 ft 4 in)
Weights: empty operating: 2,817 kg (6,210 lb); MTOW: 4,717 kg (10,400 lb)
Performance: cruise speed: 824 km/h (445 kts); range: 3,845 km (2,076 nm)
Passengers: 4-5

Agusta A.109

The A.109 is popular with air medical services

The A.109 is a twin turboshaft light/medium helicopter used for a variety of civil and military applications. Built by Italy's Agusta, it seats 6-8 passengers. In civil use it is used for emergency medical services (EMS), VIP transport, and police duties. The A.109 was designed as the Hirundo (Swallow). It made its first flight in August 1970, with production deliveries beginning in early 1976. The current production model, the A.109A Mk.II, entered service in September 1981. It uses Allison 250 engines.

The A.109 is available in numerous variants and configurations. The A.109 MAX is an improved EMS model. The A.109 POWER, announced in June 1995, will use two Pratt & Whitney Canada PW206Cs. The A.109K, designed for hot-and-high operations, is a stretched model with Turbomeca Arriel 1K1 engines and non-retractable landing gear. Now built as the A.109K2, this model is used by mountain EMS operators. The Swiss REGA mountain rescue service has ordered 15.

As of late 1995, Agusta had built over 600 A.109s, and production is continuing. For the future, Agusta is planning to develop a single turboshaft cousin for the A.109. Called the A.119 Koala, the new 8-seat machine was announced in June 1995. It could enter service in mid 1996.

Specifications (A.109 Mk.II):

Powerplant: two Allison 250-C20R/1 turboshafts, each rated at 283 kW (380 shp) maximum continuous power

Dimensions: length, rotors turning: 13.035 m (42 ft 9.25 in); height, tail rotor turning: 3.5 m (11 ft 5.75 in); width, rotor folded: 2.45 m (8 ft 0.5 in)

Weights: empty operating: 1,590 kg (3,503 lb); MTOW: 2,720 kg (5,997 lb)

Performance: cruise speed: 285 km/h (154 kts); range: 778 km (420 nm)

Passengers: 6

Bell 204/205

The 204 is known to the military as the UH-1 'Huey'

The Model 204/205 are the civil designations of Bell Helicopter's well-known UH-1 Huey military transport helicopter. Medium-sized machines, the 204/205 are used for almost every helicopter application, including medical services, cargo and personnel transport. The 204 seats 11-14 passengers, while the stretched 205 seats 15 and has an extra cabin window.

The Huey has a single T53 turboshaft powering a twin-blade main rotor. Most models have skid landing gear, but some were built with wheels and some with floats.

The first Huey was the XH-40, a demonstrator aircraft which flew in 1956. The first production UH-1A Iroquois flew in 1958. The UH-1 became synonymous with the US war in Vietnam.

Bell grew the 204/205 into the 16-seat Model 214. Bell and Agusta built about 500 of these, originally for the Iranian military, but later for other civil and military customers. The 214 first flew in 1970 and production ended in the mid 1980s. Civil and military 204/205s were also built under licence by Italy's Agusta, Japan's Fuji Heavy Industries, Taiwan's AIDC, and Germany's Dornier. The Agusta AB.204 used Rolls-Royce Gnome engines. Total production exceeded 12,000, including over 10,000 military UH-1s.

The 204/205 series was replaced by Bell's closely related 212/412 family.

Specifications (204B):

Powerplant: one Lycoming T5309A shaft-turbine rated at 820 kW (1,100 shp)
Dimensions: length, rotors turning: 17.37 m (57 ft 0 in); height, tail rotor turning: 3.81 m (12 ft 6 in); width, rotor folded (cabin): 2.39 m (7 ft 10 in)
Weights: empty operating: 2,086 kg (4,600 lb); MTOW: 3,856 kg (8,500 lb)
Performance: cruise speed: 193 km/h (120 kts); range: 530 km (330 miles)
Passengers: 8

Bell 206/407

The 206 will soon be joined by the 4-blade Model 407

The Bell 206 is a series of single and twin-turboshaft light helicopters for civil and military applications. The two major variants are the 206B JetRanger and the 206L LongRanger.
The Bell 206 began in the late 1950s. It was developed for the US Army as the OH-6, and later the OH-58A. The JetRanger, the five-seat civil variant of the series, first flew in 1962. Current production JetRanger is the 206B-3.

In 1973 Bell introduced the seven-seat LongRanger,

which features a lengthened fuselage. Current production LongRanger is the 206L-4. Bell is also widening the 206, a two-blade main rotor design, and adding a new dynamic system with a four-blade main rotor. The result is the 407, which Bell announced in February 1994. It will be built alongside the 206 series, with deliveries starting in late 1996. Most 206s are powered by a single Allison 250 turboprop. The LongRanger is also available as the 206LT TwinRanger, with two 250-C20Rs. Bell delivered the first 206LT in early 1994. The 407 will use a single 250-C47, while the 407T will use two 250-C22Bs.

Bell manufactured over 6,200 OH-6s and 206s before moving the production line to its Mirabel, Quebec, Canada plant in 1987. This line has built another 900 206s, and production is continuing.

Specifications (206B3):

Powerplant: one Allison 250-C20J turboshaft flat-rated at 236 kW (317 shp)

Dimensions: length, rotors turning: 11.82 m (38 ft 9.5 in); height, tail rotor turning: 2.91 m (9 ft 6.5 in); width, rotor folded: 1.92 m (6 ft 3.5 in)

Weights: empty operating: 737 kg (1,625 lb); MTOW: 1,451 kg (3,200 lb)

Performance: cruise speed: 214 km/h (115 kts); range: 676 km (365 nm)

Passengers: 5

Bell 212/412

The 412 is widely used by police forces

The Bell 212/412 are the follow-on models to the Bell 204/205 Huey. Like the 204/205, the 212/412 are 12-14 seat multi-role medium helicopters. The main difference between the 204/205 and 212/412 is the powerplant. The 204/205 use a single T53, while the 212/412 use a Pratt & Whitney Canada PT6T-3B Twin Pac - two PT6 turboshafts combined. The main rotor of the 212 has two blades, while the 412's has four blades. The 212 first flew in 1968 and received FAA certification in October 1970. The 412 made its

first flight in 1979 with first deliveries in January 1981.

In 1988 Bell transferred its 212 production line to its Quebec, Canada plant. The 412 line followed one year later. The 212/412 is also built under licence by Italy's Agusta.

Like the 204/205, the 212/412 has found numerous military applications. The UH-1N, used by the US Marine Corps, is a militarized Model 212. Agusta's military version is called the Griffon.

Production of both models is continuing, but most orders are for the 412. In addition to passenger transport, the series is popular with emergency medical operators, police forces and other government agencies. As of late 1995 Bell had delivered over 1,200 212/412s, and Agusta had delivered about 500 more.

Specifications (412):

Powerplant: one Pratt & Whitney Canada PT6T-3D Turbo Twin-Pac rated at 1,424 kW (1,910 shp) maximum
Dimensions: length, rotors turning: 17.12 m (56 ft 2 in); height, tail rotor turning: 4.57 m (15 ft 0 in); width, rotor folded: 2.84 m (9 ft 4 in)
Weights: empty operating: 3,018 kg (6,654 lb); MTOW: 5,397 kg (11,900 lb)
Performance: cruise speed: 230 km/h (124 kts); range: 745 km (402 nm)
Passengers: 14

Bell 222/230/430

The 4-bladed 430 is the latest version of the 222 family

The 222/230/430 is a family of twin-turboshaft seven- to ten- seat multi-role helicopters built by Bell at its Mirabel, Quebec, Canada plant. The 222 and 230 have two-blade main rotors, while the 430 has a four-blade rotor. The series is used for executive transport, emergency medical, and various utility duties.

The 222 was the first of the series. It first flew in August 1976, with first deliveries in May 1980. It was plagued by its troublesome powerplant, the Lycoming LTS101, and sales never took off. The last of 182 222s was delivered in early 1989. Bell replaced the 222 with the 230. Powered by twin Allison 250s, the 230 made its first flight in August 1991. Deliveries began in early 1993. Bell has built over 25 230s.

Bell introduced the 430 in February 1993. In addition to the new rotor, the 430 has been stretched by about 0.46 m (18 in). It also has new digital avionics, uprated 250 engines with digital controls, and an improved interior. The first 430 flew in October 1994. Deliveries will begin in January 1996. Bell is offering 430s to 230 users who trade in their current models.

Specifications (230):

Powerplant: two Allison 250-C30G2 turboshafts, each rated at 464 kW (622 shp) maximum continuous
Dimensions: length, rotors turning: 15.23 m (49 ft 11.5 in); height, tail rotor turning: 3.70 m (12 ft 1.5 in); width, rotor folded: 3.62 m (11 ft 10.5 in)
Weights: empty operating: 2,268 kg (5,000 lb); MTOW: 3,810 kg (8,400 lb)
Performance: cruise speed: 254 km/h (137 kts); range: 713 km (385 nm)
Passengers: 9

Eurocopter AS.332 Super Puma

The AS.332 is the largest Aerospatiale helicopter

The Super Puma is a large, twin-turboshaft helicopter used for both civil and military applications. Produced by the Eurocopter consortium, the Super Puma was designed by France's Aerospatiale. It first flew in September 1978, with certification and first deliveries in mid 1981.

The Super Puma is derived from the Aerospatiale/Westland SA.330 Puma built in the 1960s and 1970s and still licence-produced in Romania as the IAR-330. Compared to the Puma, the Super Puma has higher-powered engines, a bigger nose, and improved landing gear. The AS.332L,

introduced in 1983, added a stretched fuselage.

Civil applications are mostly for offshore oil and gas rig operations, but the Super Puma is also used for police, VIP, cargo and passenger transport. Major users include Bristow Helicopters and Helikopter Service. A military variant, the AS.532 Cougar, is used primarily for troop transport but also for anti-submarine and anti-ship warfare. The current production model is the AS.332L2, or Super Puma Mk II. It features advanced avionics, uprated engines, and an improved gearbox. It was certified by France's DGAC in April 1992.

As of late 1995 over 400 AS.332s have been built at Aerospatiale's Marignane plant. In addition, Indonesia's IPTN builds the type as the NAS.332L for civil and military customers.

Specifications (AS.332L2)

Powerplant: two Turbomeca Makila 1A2 turboshafts, each rated at 1,236 kW (1,657 shp) maximum continuous power
Dimensions: length, rotors turning: 19.5 m (63 ft 11 in); height, tail rotor turning: 4.97 m (16 ft 4 in); width, rotor folded: 3.86 m (12 ft 8 in)
Weights: empty: 4,660 kg (10,274 lb); MTOW: 9,150 kg (20,172 lb)
Performance: cruise speed: 851 km/h (459 kts); range: 851 km (460 nm)
Passengers: 19-24

Eurocopter AS.350 Ecureuil

The AS.350 is the single engine version of the Ecureuil

The Ecureuil (Squirrel) family comprises a series of light twin and single engine general utility helicopters. Designed and built by Eurocopter France (formerly Aerospatiale), The Ecureuil is used for emergency medical, police, cargo, VIP, and other duties. Designed as a replacement for the Aerospatiale Alouette, the AS.350 programme began in 1973. A prototype flew in June 1974, and production deliveries began in March 1978. The AS.350 is the single engine version powered by a Turbomeca Arriel.

This is marketed in North America as the A-Star. The A-Star was originally built with a Lycoming LTS101 turboshaft, but some of these have been re-engined with Arriels. The AS.355 Ecureuil 2 (Twin Star in North America) is the twin-engine variant. It first flew in September 1979. Current model is the AS.355N, certified in June 1989. The AS.355F uses two Allison 250-C20s, while the AS.355N uses two Turbomeca TM319 Arrius 1As.

Both the AS.350 and 355 are available in military versions as the AS.550/555 Fennec. These are used in every possible role including anti-tank, training, naval operations, and utility transport. Eurocopter has built over 2,000 Ecureuils, and production is continuing. About 150 of these were licence-built in Brazil by Helibras, mostly for the country's armed forces.

Specifications (AS.350B2)

Powerplant: two Turbomeca Arriel 1D1 turboshafts, each rated at 546 kW (732 shp)
Dimensions: length, rotors turning: 12.94 m (42 ft 6 in); height: 3.14 m (10 ft 4 in); width, rotor folded: 2.53 m (8 ft 3 in)
Weights: empty: 1,153 kg (2,542 lb); MTOW: 2,250 kg (4,960 lb)
Performance: cruise speed: 246 km/h (133 kts); range: 666 km (360 nm)
Passengers: 4-6

Eurocopter AS.365 Dauphin

The US Coastguard operates the HH-65A Dolphin

A medium twin-turboshaft design, the AS.365 was designed by Aerospatiale (now Eurocopter France). It is used for offshore support, emergency medical, VIP transport, pipeline patrol, and other missions. It competes with Sikorsky's S-76, but is easily distinguished by its ducted fan tail rotor, or Fenestron. The first Dauphin was the SA.360, a 10-14 seat single-engine design built by Aerospatiale in the early 1970s. The AS.365, a twin-engine version designed to replace the company's Alouette III, first flew in January 1975. The SA.360 and the early

AS.365C were powered by Astazou engines. The AS.365C was replaced by the Arriel-powered AS.365N in the early 1980s. The current Dauphin is the AS.365N2. It was certified in late 1989 and features uprated engines, new cabin doors, and a greater range of optional instrumentation.

The AS.365 is used extensively for land and naval military operations as the AS.565 Panther. The US Coast Guard operates the type as the HH-65 Dolphin. These were built by Aerospatiale in Texas and are powered by Lycoming LTS 101 engines. China's Harbin Aircraft builds the Z-9A variant under licence. As of late 1995 Eurocopter and its affiliates have built over 550 Dauphins, including 38 SA.360s and 101 HH-65As. Production is continuing.

Specifications (AS.365N2)

Powerplant: two Turbomeca Arriel 1C2 turboshafts, each rated at 471 kW (631 shp) maximum continuous power
Dimensions: length, rotors turning: 13.68 m (44 ft 11 in); height, to top of fin: 3.98 m (13 ft 1 in); width, rotor blades folded: 3.21 m (10 ft 7 in)
Weights: empty: 2,239 kg (4,936 lb); MTOW: 4,250 kg (9,370 lb)
Performance: cruise speed: 260 km/h (140 kts); range: 897 km (484 nm)
Passengers: 8-13

Eurocopter BO 105

A BO105 of Scottish Ambulance

Eurocopter's BO 105 is a five-seat twin-turboshaft light helicopter built for a variety of civil and military duties. Civil roles include sea rescue, medical services, offshore oil support, and law enforcement. The German Army has over 300 for military duties, including over 200 PAH-1 anti-tank variants.

MBB, now Eurocopter Deutschland, began BO 105 design work in 1962. A prototype flew in February 1967. The current model, BO 105CBS, was certified

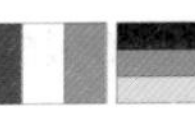

in early 1983. This version included a rear cabin stretched by 0.25 m (10 in) for extra rear seat room. The latest variant of the BO 105 is the EC Super Five. Certified in late 1993, this is a BO 105CBS with new main rotor blades, greater take-off weights, and other improvements. Many of these upgrades are derived from a recent German Army PAH-1 upgrade.

As of late 1995 over 1,200 BO 105s have been built at Eurocopter Deutschland's Donauworth facility. Also Indonesia's IPTN is building the type under licence, with over 100 NBO 105s completed since 1976. Finally, Eurocopter Canada has built over 50 of the BO 105LS model, for hot-and-high operations. BO 105 production is slowing considerably, and the type will probably be replaced by Eurocopter's EC 135 in the late 1990s.

Specifications (BO 105):

Powerplant: two Allison 250-C20B turboshafts, each rated at 298 kW (400 shp) maximum continuous
Dimensions: length, rotors turning: 11.86 m (38 ft 11 in); height, tail rotor turning: 3.02 m (9 ft 11 in); width, rotor folded: 2.53 m (8 ft 3.5 in)
Weights: empty operating: 1,301 kg (2,868 lb); MTOW: 2,500 kg (5,511 lb)
Performance: cruise speed: 240 km/h (129 kts); range: 555 km (300 nm)
Passengers: 5-6

Eurocopter EC 135

The EC135 will eventually replace the BO105

Eurocopter's EC 135 is a twin-engine 5/7-seat light helicopter designed for emergency medical, police, rescue, executive transport, and other roles. It will serve as a follow-on to the BO 105, and will be 25% less expensive to operate than the BO 105. The EC 135 began life as the BO 108, a 4-6 seat technology demonstrator created by Germany's MBB. The BO 108 first flew in October 1988. After MBB and France's Aerospatiale created Eurocopter, they modified the design with a shrouded tail rotor, the

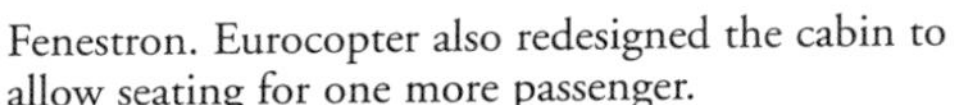

Fenestron. Eurocopter also redesigned the cabin to allow seating for one more passenger.

The EC 135 fuselage is largely composed of composite materials. It uses a four-blade bearingless main rotor and has skid landing gear. Initial models will be powered by Turbomeca Arrius turboshafts, but Pratt & Whitney Canada PW206Bs will be an option on later aircraft.

Eurocopter launched the EC 135 programme in February 1993. The first EC 135, powered by Arrius engines, made its first flight in February 1994. An EC 135 with PW206Bs flew in April 1994. Deliveries will begin in early 1996. The EC 135 will compete directly with McDonnell Douglas's MD Explorer.

Specifications (EC 135):

Powerplant: Choice of two Pratt & Whitney Canada PW206B or two Turbomeca Arrius 2B (TM 319 2R) turboshafts, each rated at 342 kW (459 shp). Data below is for aircraft with Arrius engines.

Dimensions: length, rotors turning: 12.13 m (39 ft 9.5 in); height, overall: 3.75 m (12 ft 3.5 in); width, rotor folded: 2.65 m (8 ft 8.25 in)

Weights: empty operating: 1,390 kg (3,064 lb); MTOW: 2,500 kg (5,5211 lb)

Performance: cruise speed: 261 km/h (141 kts); range: 700 km (378 nm)

Passengers: 6

Eurocopter/CATIC/Singapore Aerospace EC 120

The EC120 is a joint venture with military applications

The EC 120 is a single-engine light helicopter designed by Eurocopter, China's CATIC, and Singapore Aerospace. Aimed at the Asian market, the EC 120 will replace Gazelles, Lamas, and other aging light helicopters. The EC 120 programme was initiated by Aerospatiale's helicopter unit, now Eurocopter France. Talks began with CATIC and Australia's ASTA in 1988 for a New Light Helicopter. Australia dropped out in late 1989 and was replaced by Singapore. The three companies launched the development phase of the project in September 1991.

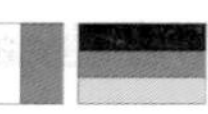

The EC 120 has a four-bladed main rotor, skid undercarriage, and a Fenestron (shrouded tail rotor). The first 300 EC 120s will use Turbomeca engines. Future EC 120s will be offered with Pratt & Whitney Canada's PW200 turboshaft. Current workshares are 61% for Eurocopter, 24% for CATIC, and 15% for Singapore Aerospace. CATIC will build the front fuselage, while Singapore will build the cockpit pedestal, doors, and tailboom, including the Fenestron.

The EC 120 made its first flight in June 1995. First deliveries will take place in September 1997. While final assembly will take place in France, there could be a second production line in China if demand warrants. There may also be a military variant for scout duties.

Specifications (EC 120)

Powerplant: one Turbomeca TM319 Arrius 1B1 turboshaft rated at 373 kW (500 shp)
Dimensions: length, rotor blades folded: 11.54 m (37 ft 10 in); height: 3.27 m (10 ft 9 in); width, including tailplane: 2.4 m (7 ft 11 in)
Weights: empty: 850 kg (1,874 lb); MTOW: 1,500 kg (3,307 lb)
Performance: cruise speed: 240 km/h (130 kts); range: 600 km (323 nm)
Passengers: 4

Eurocopter/Kawasaki BK.117

The twin-engined BK.117 is used over cities

The BK.117 is a twin-engine multi-purpose helicopter designed and built by Eurocopter Deutschland (formerly MBB) and Japan's Kawasaki Heavy Industries. There are assembly lines in both countries. Uses include executive transport, emergency medical services, offshore oil, and police operations. MBB and Kawasaki teamed up to develop the BK.117 in February 1977. Using components from MBB's BO 105 and Kawasaki's KH-7, the team built four prototypes. Kawasaki builds the fuselage, skid landing gear, and transmission. Eurocopter is responsible for systems integration and builds the main rotor head and blades, and the tail section.

The BK.117 made its first flight in June 1979. Customer deliveries began in early 1983.

The first version was the BK.117A-1. The current production model is the BK.117B-2, certified in December 1987. There is also a military variant, the BK.117M. None were ordered, but other BK.117s are used to perform military missions.

The BK.117A and BK.117B series use Textron (now AlliedSignal) Lycoming LTS 101 turboshafts, but in 1992 Eurocopter introduced the BK.117C-1, which uses Turbomeca's Arriel 1E engine. Over 470 BK.117s have been built, including over 100 from the Kawasaki production line in Gifu. Production is continuing. Indonesia's IPTN holds a licence to built the type, but has only built four. South Korea's Hyundai has also assembled about 30 BK.117s under licence.

Specifications (BK.117B-2)

Powerplant: two Lycoming LTS 101-750B-1 turboshafts, each rated at 410 kW (550 shp) maximum continuous power
Dimensions: length, rotors turning: 13 m (42 ft 8 in); height, rotors turning: 3.85 m (12 ft 8 in); width, over skids: 2.50 m (8 ft 3 in)
Weights: empty: 1,727 kg (3,807 lb); MTOW: 3,350 kg (7,385 lb)
Performance: cruise speed: 247 km/h (133 kts); range: 541 km (292 nm)
Passengers: 6-9

Kaman K-MAX

Designed as an aerial truck, the Kaman K-MAX uses intermeshing rotor blades

The K-MAX is a novel concept in civil helicopters - an aerial truck capable of lifting heavy loads. It is flown by a single pilot and carries very little internally, but it can carry up to 2,727 kg (6,000 lb) externally. It can be used for a variety of roles, including heli-logging, fire-fighting, and cargo transport. Kaman announced the K-MAX programme in March 1992, but the first prototype flew in late 1991. It was FAA

certified in August 1994, with first deliveries beginning shortly after.

While the K-MAX is a new programme, it uses some clever existing technologies which have been on the shelf for years. Most notably, it uses two intermeshing main rotor blades, developed for the HH-43 Huskie. The HH-43 was built by Kaman for the US Air Force in the 1950s. The K-MAX's single T53 turboshaft has also been in production since the late 1950s.

Kaman has built about a dozen K-MAXes, and continues to build them in small quantities. The first batch was leased to operators at $1,000 per flight hour for 1,000 flight hours per year. The K-MAX is the last design built by Kaman, so it represents the best chance for this legendary firm to stay in the helicopter business.

Specifications (K-MAX):

Powerplant: one AlliedSignal/Lycoming T53-17A-1 turboshaft rated at 1,007 kW (1,340 shp)
Dimensions: length, rotors turning: 15.85 m (52 ft 0 in); width, rotor folded 3.56 m (11 ft 8 in)
Weights: empty: 2,041 kg (4,500 lb); MTOW: 2,721 kg (6,000 lb)
Performance: cruise speed: 185 km/h (100 kts)

McDonnell Douglas MD 500

The MD500 has a distinctive egg-shaped cabin

The MD 500 is a series of single-turboshaft light helicopters built by McDonnell Douglas Helicopter Systems (MDHS) for numerous civil and military applications. The MD 500 has a five-blade main rotor, an egg-shaped cabin and skid landing gear. While it has a conventional tail rotor, the MD 500 is closely related to the MD 520N and 630N. The MD 500 was originally the OH-6 Cayuse, built by Hughes (later MDHS) for the US Army. The first

OH-6 flew in February 1963. In April 1965 Hughes decided to develop the Model 500 civil variant, with production beginning in November 1968.

The MD 500 is available as the MD 500 Defender, an armed variant for military applications. The MD 500E civil variant has a de-rated Allison 250-C20B turboshaft. The MD 530F variant features a 250-C30, with better take-off performance.

As of late 1995 MDHS and Hughes had built over 3,900 MD 500s, including over 1,400 OH-6s. Big civil users include police departments and other government agencies. Italy's Agusta and Japan's Kawasaki build the type under licence. Korean Air built over 300 under licence between 1976 and 1988. Production is continuing at MDHS, Agusta and Kawasaki.

Specifications (MD500E):

Powerplant: one Allison 250-C20B rated at 261 kW (350 shp) maximum continuous power
Dimensions: length, rotors turning: 8.61 m (28 ft 3 in); height, tail rotor turning: 2.67 m (8 ft 9 in); width, rotor folded: 1.91 m (6 ft 3 in)
Weights: empty operating: 655 kg (1,445 lb); MTOW: 1,361 kg (3,000 lb)
Performance: cruise speed: 245 km/h (132 kts); range: 431 km (233 nm)
Passengers: 4

McDonnell Douglas MD 520N/630N

The MD520N is the first helicopter to use the NOTAR system

The MD 520N and 630N are two no-tail rotor (NOTAR) variants of the MD 500 single-engine light helicopter family. The 520N uses the MD 500 cabin, seating 2-4 passengers. The 630N uses a stretched cabin, seating 7-8 passengers. Both types have a graphite composite tailboom with an 'H' tail, housing the NOTAR exhaust duct.

McDonnell developed the NOTAR system under a US Army contract, although the Army has declined

to use the system. McDonnell announced the 520N model for the civil market in February 1988. FAA certification was granted in September 1991. Original plans called for a 530N hot-and-high variant, but the 520N had sufficient engine power for most operators and the 530N was dropped.

The MD 630N was first revealed in January 1995, but a prototype had flown in November 1994. It has a sixth main rotor blade and an uprated Allison engine. Arizona's AirStar Helicopter provided the launch order in February 1995, and the first 630N will be delivered in late 1996. Meanwhile, MDHS is continuing production of the MD 520N. By the end of 1995 over 80 had been delivered. They are especially popular for police duties - the Los Angeles County Sheriff's Office has ordered nine.

Specifications (McDonnell Douglas MD520N):

Powerplant: one Allison 250-C20R turboshaft de-rated to 280 kW (375 shp) maximum continuous power
Dimensions: length, rotors turning: 9.78 m (32 ft 1.25 in); height, tail rotor turning: 2.74 m (9 ft 0 in); width, rotor folded: 2.01 m (6 ft 7.25 in)
Weights: empty operating: 742 kg (1,636 lb); MTOW: 1,519 kg (3,350 lb)
Performance: cruise speed: 249 km/h (135 kts); range: 402 km (217 nm)
Passengers: 4

McDonnell Douglas MD Explorer

The MD Explorer awaits possible military orders

The MD Explorer is a twin-engine civil helicopter designed for offshore oil support, emergency medical, law enforcement, and other applications. It is the first all-new helicopter to feature a No Tail Rotor (NOTAR) system, developed by McDonnell Douglas for its (losing) entry to the US Army light helicopter competition. It also features advanced digital avionics.

The MD Explorer began life in 1986 as a market study for a new advanced 8-seat helicopter. Designated MDX, the programme was launched in January 1989. Later the same year, Australia's Hawker de Havilland and Japan's Kawasaki signed on as partners. Hawker is building fuselages and tailbooms, while Kawasaki is providing transmissions.

Pratt & Whitney Canada also signed on as a partner, but the company's PW206 engine will only be exclusive on the first 128 production aircraft. After that, the MD Explorer will also be available with Turbomeca's TM319-2C Arrius engine.

The MD Explorer made its first flight in December 1992. It received FAA certification in December 1994. The same month, the first production aircraft was delivered to Petroleum Helicopters Inc. Later that month, a second machine was delivered to Rocky Mountain Helicopters. This aircraft had a full aeromedical interior.

For the future, McDonnell Douglas is considering a military version of the MD Explorer. It is also studying a single-engine version, with an LHTEC CTS800.

Specifications (MD Explorer)

Powerplant: two Pratt & Whitney Canada PW206B turboshafts, each rated at 469 kW (629 shp) take-off
Dimensions: length, rotors turning: 11.99 m (39 ft 4 in); height: 3.66 m (12 ft); fuselage width: 1.63 m (5 ft 4 in)
Weights: empty: 1,458 kg (3,215 lb); MTOW: 2,699 kg (5,950 lb)
Performance: cruise speed: 274 km/h (148 kts); range: 555 km (299 nm)
Passengers: 8

Sikorsky S-61

A few S-61s are used for offshore transport in the UK

The S-61 is a large twin-engine transport helicopter capable of seating up to 30 passengers. While primarily built as a naval and military design (designated SH-3 Sea King and CH-3), the S-61 was also the first widely used helicopter airliner. An S-61 prototype first flew in December 1960, although a military SH-3 had flown in March 1959. FAA certification was awarded in November 1961.

There were two basic versions of the S-61. The S-61L uses retractable wheel landing gear for land operations. It first flew in 1960. The S-61N is an

amphibious version with stabilizing floats and a sealed hull. Both types have a small retractable nose radome carrying a weather radar and are powered by two General Electric CT58 turboshafts turning a five-blade main rotor.

Sikorsky built 136 S-61s for civil users, with production ending in 1980. Additional SH-3s were converted to civilian use. Many new-build S-61s went to British International Helicopters, which still operates 15 S-61Ns. Several other operators use the S-61, including Norway's Helikopter Service, which operates 16 S-61Ns. Italy's Agusta also built about five examples of the AS-61N1 Silver, a version of the S-61N with rearranged windows and smaller sponsons. This variant, designed for offshore and search-and-rescue operations, first flew in July 1984.

Specifications (S-61N):

Powerplant: two General Electric CT58-140-2 turboshaft engines, each rated at 1,118 kW (1,500 shp)
Dimensions: length, rotors turning: 22.2 m (72 ft 10 in); height, cabin: 1.92 m (6 ft 3.5 in); width: cabin: 1.98 m (6 ft 6 in)
Weights: empty operating: 6,010 kg (13,255 lb); MTOW: 8,620 kg (19,000 lb)
Performance: cruise speed: 222 km/h (120 kts); range: 833 km (450 nm)
Passengers: 30

Sikorsky S-76

The S-76 is Sikorsky's only dedicated civil helicopter

The Sikorsky S-76 is a twin-turboshaft transport and utility helicopter used for VIP, emergency medical, offshore oil rig support and other missions. A medium-sized machine, transport variants of the S-76 seat 8-12 passengers. It competes with Eurocopter's AS.365. Sikorsky first announced the S-76 in early 1975. It was developed with the company's popular UH-60 Black Hawk military helicopter and uses related technologies. The first of four S-76 prototypes flew in March 1977. The S-76 is available in a bewildering variety of models with different engines

on each. First was the S-76 and S-76 Mark II both powered by Allison 250 turboshafts. Then came the heavier S-76B with Pratt & Whitney Canada PT6B-36 engines. This was certified in 1985. The latest version is the S-76C. Using the S-76B airframe and powered by Turbomeca Arriels, the S-76C replaced the S-76A+. It was introduced at the 1989 Paris Air Show, and deliveries began in mid 1991. By early 1996, it will be replaced by the S-76C+, with uprated Arriel engines. Some Allison-powered S-76s were retrofitted with Arriels, becoming S-76A+s.

Sikorsky also proposed a military variant, the H-76. This remains unlaunched, but Japan's navy uses the S-76C for search and rescue operations, and other militaries use the type for various duties as well. As of late 1995 Sikorsky had delivered over 400 S-76s.

Specifications (S-76C)

Powerplant: two Turbomeca Arriel 1S1 turboshafts, each rated at 539 kW (723 shp) maximum continuous power
Dimensions: length, rotors turning: 16 m (52 ft 6 in); height, tail rotor turning: 4.41 m (14 ft 6 in); fuselage width: 2.13 m (7 ft)
Weights: empty: 2,849 kg (6,282 lb); MTOW: 5,171 kg (11,400 lb)
Performance: cruise speed: 269 km/h (145 kts); range: 789 km (430 nm)
Passengers: 8-12

Glossary of Aviation Terms

AIDC	Aero Industry Development Center, a Taiwanese aircraft manufacturing company
Ailerons	control surfaces, usually found on trailing edge of outer wing
AIR	Aero International Regional,
AMR	Parent corporation of American Airlines
ASTA	Aerospace Technologies of Australia, an aircraft manufacturing company
ATR	Avions de Transport Regional, a regional aircraft manufacturing alliance between Alenia and Aerospatiale
BAe	British Aerospace
CAA	Civil Aviation Authority, the UK government agency responsible for civil aviation
CASA	Construcciones Aeronauticas SA, a Spanish aircraft manufacturer
CATIA	Conception assistée tridimensionelle interactive d'applications, a French computer-aided aircraft design system
CATIC	China National Aero Technology Industrial Corporation
De-rated engine	an engine with a maximum power output set below its maximum potential level
DGAC	Direction General a l'Aviation Civile, France's government aviation agency
DHC	De Havilland Canada
DHL	An international freight delivery service
EFIS	Electronic Flight Instrumentation System: digital

	cockpit displays, also known as a 'glass cockpit'
FAA	Federal Aviation Administration, the US government agency responsible for aviation
Fenestron	a shrouded fan used as a helicopter tail rotor
Flaps	Movable control surface designed to increase lift, found on trailing wing edge
Foreplanes	Horizontal foreplanes mounted forward of the wings, designed to improve low-speed and take-off performance
Hot-and-High operations	Flights in hotter than usual airfields and/or from airfields considerably higher than sea level (both conditions requiring more engine power)
Hush-kitting	Installation of devices which make jet engines quieter, usually at the expense of performance and/or fuel consumption
ILFC	International Lease Finance Corporation, a US aircraft leasing company
Intercontinental range	sufficient to cross the Atlantic Ocean or greater distances, at least 5920 km (3200 nm)
IPTN	Industri Pesawat Terbang Nusantra, an Indonesian aircraft manufacturer
JAA	Joint Airworthiness Authorities, the European aviation authority
KLM	Koninklijke Luchtvaart Maatschappij, flag carrier of the Netherlands

kts (knots)	nautical miles per hour
lbst (pounds thrust)	a measure of turbofan engine power
MBB	Messerschmitt Bolkow Blohm, a German aircraft design firm, now part of Daimler Benz Aerospace
MTOW	Maximum Take Off Weight
MTU	Motoren und Turbinen-Union, a German aircraft manufacturer
Nacelles	Streamlined pods designed to house aircraft engines
NAPA	Novosibirsk Aircraft Production Association
nm	nautical miles
podded engine	engines in housings, usually nacelles (q.v.)
Radome	Protective covering for an aircraft's radar, usually found at the front tip of the fuselage
Rear-facing pusher configuration	A turbo-prop aircraft with engines facing aft, and propellers behind the engine and wing
SAIC	Shanghai Aviation Industrial Corporation, a Chinese aircraft manufacturer
SAS	Scandanavian Airlines System
Sensor platform	an aircraft designed primarily to carry a sensor such as maritime radar
shp	shaft horse power
skid landing gear	helicopter landing equipment, an alternative to wheels
skid undercarriage	see above

sponsons	projections from an amphibious aircraft hull desinged to provide stability in the water
TAT	European Airlines, a French carrier
TNT wing	Dornier's Tragflugels Neuer Technologie, or new technology wing
Transcontinental range	sufficient to cross North America, i.e. about 4810 km (2600 nm)
Trunkliner	Airliner designed for 'trunk' routes between major cities within a country (usually a narrowbody carrying 150-200 seats)
Turbofan	Jet engine with a large fan in front, generating most of the thrust from air which bypasses the central engine core: the primary propulsion system for airliners and business jets
Turboprop	Jet engine geared to an external propeller, usually found on regional aircraft
Turboshaft	Jet engine coupled to a shaft, usually to turn helicopter rotor blades
UPS	United Parcel Service, a US cargo carrier
ventral delta fins	fixed or movable fins on the underside of the fuselage, often at the tail
wing fairings	secondary structures in front of the wing, next to the fuselage, designed to reduce drag
winglets	upturned wingtips designed to improve wing cruise efficiency
wingtip fences	smaller winglets, often turned upward and downward from the wing

COLLINS GEM
BABIES'
names
a
?
z
a mine of information

COLLINS GEM
BEER
a mine of information

COLLINS GEM
BIRDS
a mine of information

COLLINS GEM
CALORIE
Counter
a mine of information

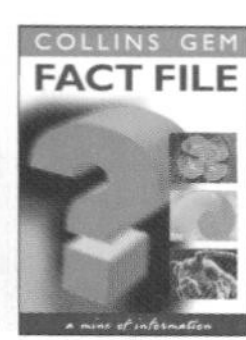
COLLINS GEM
FACT FILE
a mine of information

COLLINS GEM
FENG SHUI
a mine of information

COLLINS GEM
FLAGS
a mine of information

COLLINS GEM
Healthy
EATING
a mine of information

COLLINS GEM
QUOTATIONS
a mine of information

COLLINS GEM
SAS
Self-Defence
a mine of information

COLLINS GEM
SAS
Survival Guide
a mine of information

COLLINS GEM
SEASHORE
a mine of information

COLLINS GEM
TREES
a mine of information

COLLINS GEM
Understanding
DREAMS
a mine of information

COLLINS GEM
WILD
flowers
a mine of information

COLLINS GEM
WINE
Dictionary
a mine of information